Renewable Energy Integration to the Grid

Renewable Energy Integration to the Grid
A Probabilistic Perspective

Edited by
Neeraj Gupta
Anuradha Tomar
B Rajanarayan Prusty
Pankaj Gupta

CRC Press is an imprint of the
Taylor & Francis Group, an **informa** business

MATLAB® and Simulink® are trademarks of The MathWorks, Inc. and are used with permission. The MathWorks does not warrant the accuracy of the text or exercises in this book. This book's use or discussion of MATLAB® and Simulink® software or related products does not constitute endorsement or sponsorship by The MathWorks of a particular pedagogical approach or particular use of the MATLAB® and Simulink® software.

First edition published 2022
by CRC Press
6000 Broken Sound Parkway NW, Suite 300, Boca Raton, FL 33487-2742

and by CRC Press
2 Park Square, Milton Park, Abingdon, Oxon, OX14 4RN

© 2022 selection and editorial matter, Neeraj Gupta, Anuradha Tomar, B Rajanarayan Prusty and Pankaj Gupta; individual chapters, the contributors

CRC Press is an imprint of Taylor & Francis Group, LLC

Reasonable efforts have been made to publish reliable data and information, but the author and publisher cannot assume responsibility for the validity of all materials or the consequences of their use. The authors and publishers have attempted to trace the copyright holders of all material reproduced in this publication and apologize to copyright holders if permission to publish in this form has not been obtained. If any copyright material has not been acknowledged please write and let us know so we may rectify in any future reprint.

Except as permitted under U.S. Copyright Law, no part of this book may be reprinted, reproduced, transmitted, or utilized in any form by any electronic, mechanical, or other means, now known or hereafter invented, including photocopying, microfilming, and recording, or in any information storage or retrieval system, without written permission from the publishers.

For permission to photocopy or use material electronically from this work, access www.copyright. com or contact the Copyright Clearance Center, Inc. (CCC), 222 Rosewood Drive, Danvers, MA 01923, 978-750-8400. For works that are not available on CCC please contact mpkbookspermissions@ tandf.co.uk

Trademark notice: Product or corporate names may be trademarks or registered trademarks and are used only for identification and explanation without intent to infringe.

ISBN: 978-0-367-74794-7 (hbk)
ISBN: 978-1-032-22264-6 (pbk)
ISBN: 978-1-003-27185-7 (ebk)

DOI: 10.1201/9781003271857

Typeset in Times
by codeMantra

Contents

Preface vii
Editors ix
Contributors xi

Chapter 1 Renewable Energy Scenario of the World and Future Pattern 1

Karan Singh Joshal, Ashiq Hussain Lone, Neeraj Gupta,
Anuradha Tomar, and Rakesh Sehgal

Chapter 2 Technical Challenges in Renewable Generations
and Their Integration to Grid 29

D. Blandina Miracle, Rajkumar Viral,
Pyare Mohan Tiwari, and Mohit Bansal

Chapter 3 Renewable Energy Integration Issues from Consumer
and Utility Perspective 49

K. Pritam Satsangi, G. S. Sailesh Babu,
Bhagwan Das Devulapalli, and Ajay Kumar Saxena

Chapter 4 Voltage Issues in Power Networks with Renewable
Power Generation 79

Sushil Kumar Gupta and Kapil Gandhi

Chapter 5 Reactive Power Management in Power Systems
Integrated with Renewable Generations 107

Satish Kumar and Ashwani Kumar

Chapter 6 Optimum Scheduling and Dispatch of Power Systems
with Renewable Integration 131

Abhishek Rajan and Bimal Kumar Dora

Chapter 7 Role of Stochastic Optimization for Power System
Operation and Decision-Making 165

Wen-Shan Tan and Mohamed Abdel Moneim Shaaban

Chapter 8 Dependence Modeling of Multisite Renewable Generations 189

B Rajanarayan Prusty and Satyabrata Das

v

Chapter 9 Probabilistic Steady-State Analysis of Power Systems Integrated with Renewable Generations...199

Vikas Singh, Tukaram Moger, and Debashisha Jena

Chapter 10 Risk Evaluation of Electricity Systems with Large Penetration of Renewable Generations...239

Sheng Wang, Lalit Goel, and Yi Ding

Index...261

Preface

In recent days, there is a globally increasing interest in integrating renewable generations into the power systems. These generations are intermittent and highly volatile; their integration into power systems dramatically affects the power system variables, a significant concern in power system studies. The application of data analytics is crucial in these studies. Hence, a book focusing on the various aspects of power systems with integration of such renewable sources is the need of the hour. This book is expected to instill in final-year graduate and postgraduate novice researchers the research interest on "application of probabilistic methods to power systems integrated with renewable generations." In addition to the academicians, this book adds a new dimension for the engineers in the power industry to ascertain the level of renewable penetration at a given location from system reliability in terms of probability values. A total of ten chapters are planned to cover in detail the various issues emanated from renewable generation-rich power systems. Special attention is given to adopting probabilistic approaches for uncertainty quantification, dependence modeling, steady-state analysis, and risk assessment. Each chapter's content is briefly elucidated underneath.

Chapter 1 elaborates on renewable technologies, considering their present and future scenarios, working operation, environmental impacts, and cost analysis. The fact that renewable energy sources have full potential to change their status from being a supplementary source to the primary source of energy-producing technology is indicated.

Chapter 2 examines alternative renewable energy sources and thus evaluates the different controllers, fault types, and different solutions. Moreover, the technical challenges with the installations of renewable generations, case studies of mitigation of technical challenges, problems associated with integration at transmission and distribution voltage level, and possible solutions to address renewable integration challenges are discussed.

Chapter 3 emphasizes the importance of performance analysis and its parameters, as per standards, for a solar photovoltaic microgrid. Further, it applies the new standard, IEC 61724: 2017, on the 40 kWp microgrid in Agra, India.

Chapter 4 briefs on various voltage issues occurring in the power system operation. Further, the impact of renewable energy integration on different prospects is discussed in detail.

Chapter 5 presents the importance and management of reactive power with renewable energy resources in the present market model condition. Some conventional methods for reactive power management and assessment are also explained. Critical issues with reactive power management like voltage stability are addressed.

Chapter 6 is designed to shed light on various methods, practical constraints, and challenges for the optimum dispatch of active and reactive powers both in the absence and in the presence of renewable generations. Also, it emphasizes the modeling of the intermittent nature of renewable generations and uncertainty in load demand.

Chapter 7 covers the state-of-the-art stochastic optimization techniques and their applications in power system operation, emphasizing generation rescheduling. Stochastic optimization principles are introduced, and the generation scheduling problem amenable to stochastic optimization is presented. Further, stochastic optimization categories, including robust optimization and chance-constrained programming, are deliberated.

Chapter 8 provides a theoretical detail of various copula-based dependence modeling steps for probabilistic power system analysis. The advantage of the copula function compared to conventional Pearson's product-moment correlation coefficient in modeling the dependence in the part of the distribution where the association is potent is deliberated comprehensively.

Chapter 9 presents the probabilistic assessment of renewable energy–integrated power systems to analyze various uncertainty issues in transmission systems. Probabilistic load flow methods, such as Monte-Carlo simulation, cumulant method, and point estimation method, are explored by applying them to a sample 10-bus test system.

Chapter 10 introduces the importance of spatial-temporal risk evaluation in power systems. The time-space Markov modeling technique is proposed to characterize the stochastic process of wind power and the dependencies over various locations. The operational reliabilities of gas-fired generators and power-to-gas facilities are modeled in a multistage manner. The concept of traditional over-limit risk indices is extended to quantify the spatial-temporal risk of the electricity system.

MATLAB® is a registered trademark of The MathWorks, Inc. For product information, please contact:

The MathWorks, Inc.
3 Apple Hill Drive
Natick, MA 01760-2098 USA
Tel: 508-647-7000
Fax: 508-647-7001
E-mail: info@mathworks.com
Web: www.mathworks.com

Editors

Neeraj Gupta, Ph.D., was born in Jammu, India in 1982. He received the B.E. degree in Electrical Engineering from Jammu University, Jammu, India in 2005, the M.E. degree in power system and electric drives from Thapar University, Patiala, India in 2008, and the Ph.D. degree in power systems from the Indian Institute of Technology Roorkee, Roorkee, India in 2015. He was a faculty with the Electrical and Instrumentation Engineering Department, Thapar University, from 2008 to 2009 and Adani Institute of Infrastructure Engineering, Ahmedabad, India in 2015. Since 2015, he has been working as a faculty with the Electrical Engineering Department, National Institute of Technology, Hamirpur, India up to 2018. Presently he is working as an Assistant Professor in National Institute of Technology, Srinagar, India. He is a senior member of IEEE and reviewer of all the reputed journals of power systems in publishers like IEEE, Elsevier, Taylor and Francis, Wiley, IET, etc. His research interests include uncertainty quantification of power system, probabilistic power system, solar, wind, and electric vehicle technologies.

Anuradha Tomar has 12 years of experience in research and academics. She is currently working as an Assistant Professor in Instrumentation & Control Engineering Division of Netaji Subhas University, Delhi, India. Dr. Tomar has completed her postdoctoral research in Electrical Energy Systems Group in Eindhoven University of Technology (TU/e), the Netherlands and has successfully completed European Commission's Horizon 2020, UNITED GRID and UNICORN TKI Urban Research projects. She has received her B.E. degree in Electronics Instrumentation & Control with Honours in the year 2007 from the University of Rajasthan, India. In the year 2009, she has completed her M.Tech. degree with Honours in Power System from the National Institute of Technology Hamirpur. She has received her Ph.D. in Electrical Engineering from the Indian Institute of Technology Delhi (IITD). Dr. Anuradha Tomar has committed her research work efforts toward the development of sustainable, energy efficient solutions for the empowerment of society and humankind. Her areas of research interest are Operation & Control of Microgrids, Photovoltaic Systems, Renewable Energy–Based Rural Electrification, Congestion Management in LV Distribution Systems, Artificial Intelligent & Machine Learning Applications in Power System, Energy Conservation and Automation. She has authored or coauthored 69 research/review papers in various reputed international and national journals and conferences. She is an editor for books with international publications like Springer and Elsevier. Her research interests include photovoltaic systems, microgrids, energy conservation, and automation. She has also filled seven Indian patents on her name. Dr. Tomar is a senior member of IEEE and a life member of ISTE, IETE, IEI, and IAENG.

B Rajanarayan Prusty (Senior Member, IEEE) is presently working as an Assistant Professor (Sr. Grade) in the School of Electrical Engineering, Vellore Institute of Technology (VIT), Vellore. He has obtained his Ph.D. from the National Institute of

ix

Technology Karnataka (NITK), Surathkal. He successfully delivers quality education to the student community by developing a student-friendly environment. His exceptional research work during Ph.D. work has led him to crown the prestigious POSOCO Power System Awards (PPSA) for 2019 under the doctoral category by Power System Operation Corporation Limited in partnership with FITT, IIT Delhi. He has 15 SCI journal publications and 30 IEEE conference publications to his credit. In recognition of his research publications from 2017 to 2019, he is awarded the University Foundation day Research Award-2019 from BPUT, Odisha, INDIA. He has authored seven book chapters published in Elsevier and Springer. He has coauthored a textbook entitled *Power System Analysis: Operation and Control* in I. K. International Publishing House Pvt. Ltd., New Delhi, ISBN 9789382332954. He has been an active reviewer since 2015 and has reviewed 200 manuscripts submitted to reputed SCI-indexed journals/conferences. Presently he is the Associate Editor (Electric Power Engineering) of *Journal of Electrical Engineering & Technology*, Springer. His research interest includes time series preprocessing and forecasting, high-dimensional dependence modeling, and probabilistic power system analyses.

Pankaj Gupta received a B.E. degree in Electrical Engineering from Bhilai Institute of Technology, Pt. Ravi Shankar Shukla University, Raipur, India in 2000 and M.E. degree in Electrical Engineering from Delhi College of Engineering, Delhi University, Delhi, India in 2003. He obtained a Ph.D. degree in Electrical Engineering from NIT, Kurukshetra, India in 2017. He is conferred with prestigious POSOCO Power System Award-2017 for outstanding PhD research work entitled "Protection Issues of Grid Connected Distributed Generation" by the Power System Operation Corporation Limited, a subsidiary of Power Grid Corporation, India in partnership with Foundation for Innovation and Technology Transfer, IIT, Delhi. He is working with Indira Gandhi Delhi Technical University for Women, Delhi, India as an Assistant Professor since 2005. His research interests include power system protection, microgrid control and protection, smart grid technologies, and islanding detection techniques.

Contributors

Mohit Bansal
Department of Electrical and
 Electronics Engineering
G L Bajaj Institute of Technology &
 Management, India

Bhagwan Das Devulapalli
Department of Electrical Engineering
Dayalbagh Educational Institute
Agra, India

Satyabrata Das
Department of Electronics and
 Communication Engineering
NIST Rourkela
India

Yi Ding
College of electrical engineering
Zhejiang University
Hangzhou, China

Bimal Kumar Dora
Department of Electrical and
 Electronics Engineering
National Institute of Technology Sikkim
Ravangla, India

Kapil Gandhi
Department of Electrical and
 Electronics Engineering
KIET Group of Institutions
Delhi NCR, India

Lalit Goel
Department of Electrical Engineering
 School
Nanyang Technological University
Singapore, Singapore

Sushil Kumar Gupta
Department of Electrical Engineering
Deenbandhu Chhotu Ram University of
 Science and Technology
Murthal, India

Debashisha Jena
Department of Electrical and
 Electronics Engineering
National Institute of Technology
 Karnataka
Surathkal, India

Karan Singh Joshal
Department of Electrical Engineering
National Institute of Technology
Srinagar, India

Ashwani Kumar
Department of Electrical and
 Electronics Engineering
National Institute of Technology
Kurukshetra, India

Satish Kumar
Department of Electrical and
 Electronics Engineering
KIET Group of Institutions
Ghaziabad, India

Ashiq Hussain Lone
Department of Electrical Engineering
National Institute of Technology
Srinagar, India

D. Blandina Miracle
Department of Electrical and
 Electronics Engineering
Amity University
India

Tukaram Moger
Department of Electrical and
  Electronics Engineering
National Institute of Technology
  Karnataka
Surathkal, India

K. Pritam Satsangi
Department of Electrical Engineering
Dayalbagh Educational Institute
Agra, India

Abhishek Rajan
Department of Electrical and
  Electronics Engineering
National Institute of Technology Sikkim
Ravangla, India

G. S. Sailesh Babu
Department of Electrical Engineering
Dayalbagh Educational Institute
Agra, India

Ajay Kumar Saxena
Department of Electrical Engineering
Dayalbagh Educational Institute
Agra, India

Mohamed Abdel Moneim Shaaban
Department of Electrical Engineering
Universiti Malaya
Kuala Lumpur, Malaysia

Vikas Singh
Department of Electrical and
  Electronics Engineering
National Institute of Technology
  Karnataka
Surathkal, India

Rakesh Sehgal
Department of Mechanical Engineering
National Institute of Technology
Srinagar, India

Wen-Shan Tan
School of Engineering
Monash University Malaysia
Subang Jaya, Malaysia

Pyare Mohan Tiwari
Department of Electrical and
  Electronics Engineering
Amity University
India

Rajkumar Viral
Department of Electrical and
  Electronics Engineering
Amity University
India

Sheng Wang
State Grid (Suzhou) City & Energy
  Research Institute Co., Ltd.
China

1 Renewable Energy Scenario of the World and Future Pattern

Karan Singh Joshal[a], Ashiq Hussain Lone[a], Neeraj Gupta[a], Anuradha Tomar[b], and Rakesh Sehgal[a]

[a] National Institute of Technology Srinagar, Jammu and Kashmir

[b] Netaji Subhas University of Technology, Delhi

CONTENTS

1.1 Introduction 2
    1.1.1 Stated Policies 3
    1.1.2 Sustainable Development 3
1.2 Future Patterns of Renewable Sources 4
    1.2.1 The Future of Wind 4
    1.2.2 Wind Power Is Expected to Reach 43 GW by 2030 5
    1.2.3 Future Prospects for Solar 6
1.3 Renewable Energy Technologies and Ongoing Developments 7
1.4 Wind Energy 7
    1.4.1 Basic Principle 7
    1.4.2 Technical Aspects of Wind Turbines 8
        1.4.2.1 Rotor 8
        1.4.2.2 Nacelle 9
        1.4.2.3 Gearbox 9
        1.4.2.4 Generator 9
        1.4.2.5 Tower and Foundation of Wind Turbine 10
    1.4.3 Control Technique Requirements 10
        1.4.3.1 Pitch Adjustment Control Method 10
        1.4.3.2 Yaw Mechanism 10
    1.4.4 Economical Analysis and Environmental Impacts 11
        1.4.4.1 Economic Analysis 11
        1.4.4.2 Environmental Impacts 11
1.5 Solar Energy 11
    1.5.1 Solar Power Technology 12
    1.5.2 CSP Systems 12
    1.5.3 PV Systems 12

DOI: 10.1201/9781003271857-1

	1.5.4	Environmental Impacts	14
	1.5.5	Economic Analysis of Solar Energy	14
1.6	Biomass Energy		15
	1.6.1	Biomass: Types and Harvesting	15
	1.6.2	Bioenergy Conversion Technologies	16
	1.6.3	Environmental Impacts of Biomass Energy	17
	1.6.4	Economic Analysis of Biomass	18
1.7	Ocean Energy		18
	1.7.1	Different Types of Ocean Thermal Technologies	18
	1.7.2	Environmental Impacts	20
	1.7.3	Economic Analysis of Ocean Energy Technology	20
1.8	Geothermal Energy		21
	1.8.1	Resources of Geothermal Energy	21
	1.8.2	Geothermal Energy Technologies	22
		1.8.2.1 Dry Steam Power Plants	22
		1.8.2.2 Flash Point Power Plants	22
		1.8.2.3 Binary Cycle Power Plant	24
	1.8.3	Cost Analysis	24
	1.8.4	Environmental Impacts	24
1.9	Concluding Remarks		24
References			26

1.1 INTRODUCTION

Renewable energy is the energy that can be produced again and again within a period of time. The main resources of renewable energy are photovoltaic (PV), wind, biogas, hydro, biomass, tidal, geothermal, etc. The use of these resources causes the least amount of pollution, and hence the dependency on them is increasing day by day. The region-wise analysis and future pattern of these energy resources is explained in detail.

The growth of renewable energy technologies has escalated in recent years, with the power sector playing a significant role, which is related to the low costs of solar PV and wind power. Renewable energy usage, on the other hand, is also slow in end-use sectors like manufacturing and construction.

Since 2018, the generation of electricity from renewable resources has shown significant growth (Abbaszadeh et al. 2013). The power output in 2018 was about 450 Terawatt-hours (TWh), or it was up by 7% with regard to the previous years, which is about quarter times higher than total power generation. The production of power from renewable sources such as PV, wind, and hydro has shown a 90% increase. In 2018, about 180 GigaWatts (GW) of renewable energy was produced. Now International Energy Agency has estimated that from 2021 onwards, there will be an annual increase in the capacity.

Here reduction in the cost by renewables and advancement in the technology are providing large opportunities for the transitions in energy. The renewable energy sources will provide almost half of the energy by 2040 as stated in Stated Policies Scenario and Sustainable Development Scenario.

Renewable Energy Scenario of the World

1.1.1 STATED POLICIES

As per Stated Policies Scenario, the use of renewable sources (except biomass) will increase from 990 Mtoe to approximately 2,260 Mtoe in 2040. According to the Sustainable Development Scenario, by 2040, wind energy (8,300 TWh) and solar PV energy (7,200 TWh) are forecast to overtake hydropower (6,950 TWh). The energy consumption in transport sector is expected to boost to 600 Mtoe by using biofuels, which may lead to about 60% increase in using biofuels without considering the electricity consumed by electric vehicles.

According to the Stated Policies Scenario, about 8,500 GW of new energy will be included worldwide within 2040, and two-thirds of this will be from renewables. In general, maximum of a region's renewable have accounted for most of the capacity additions. Renewables have increased by 80% in the European Union and China, and they supply more than half of the electricity, but in Southeast Asia and the Middle East, they provide less than half of the electricity. Solar energy accounts for the majority of the energy in the majority of the regions, including India and China. This is shown in Figure 1.1a and b where series 1 represents the year 2018, series 2 represents the year 2030 and series 3 represents the year 2040. Figure 1.1a shows the generation of electricity from renewable resources by province and scenario, 2018–2040 in TWh Stated Policy Scenario, and Figure 1.1b shows electricity from capability of renewable resources by province and scenario, 2018–2040 in GW Stated Policies Scenario.

1.1.2 SUSTAINABLE DEVELOPMENT

As per the Sustainable Development Scenario, 80% of the energy is from renewables. The investment in renewable energy slightly declined in 2018 to about $390 billion, but the investment in this field continued to show growth. The investment in renewable will accelerate whatsoever the scenario world will follow.

According to the Stated Policies Scenario, the venture in renewables reached a growing sum involving at the moment and after 2040 about $10 trillion. The investment in Sustainable Development Scenario increases at a quicker pace which reflects the stronger plan support and the job of center.

According to the Specified Policies Scenario, renewable energy assets are expected to rise to $10 trillion between now and 2040. In the Sustainable Development Scenario, speculation rises more quickly, indicating a more effective approach, as well as the position of the internal mission and the critical task, with the aim of these new power technologies contributing to the achievement of sustainable energy priorities. Figure 1.2a shows the amount of renewable electricity produced by province and scenario from 2018 to 2040 in GW Sustainable Development Scenario, and Figure 1.2b represents generation from renewable electricity by province and scenario sustainable development, 2018–2040 in TWh Sustainable Development Scenario.

The annual investment on the renewable and future investment in billion USD is shown in Table 1.1.

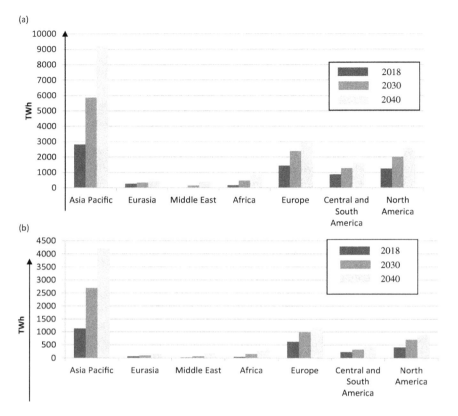

FIGURE 1.1 (a) Renewable electricity generation by region and scenario, stated policies 2018–2040 in TWh Stated Policy Scenario (Gielen et al. 2019). (b) Capabilities of renewable resources by province and scenario, 2018–2040 in GW Stated Policies Scenario (Gielen et al. 2019).

1.2 FUTURE PATTERNS OF RENEWABLE SOURCES

All renewable energy technologies, including wind and solar, have the highest potential for development, making them the scalability champions. In future, these innovations will lead the way in India's renewable energy market.

The solar and wind energy industries will account for the majority of this development. As previously mentioned, a variety of technological and market advancements are projected to boost the growth of renewable energy in the future. The influence of these advances on the potential of the wind and solar energy sectors is briefly described in this section.

1.2.1 THE FUTURE OF WIND

The wind turbines are available in the range of 2–3 MW. Increased turbine size has also made wind turbine construction in 'tier 3' areas more cost-effective. Previously, only class 2 and class 1 sites were considered for wind farm growth. The term "class" refers to how sites are categorized based on the wind speeds available. Wind speeds that are higher result in more energy being produced. Wind speeds are superior at

Renewable Energy Scenario of the World

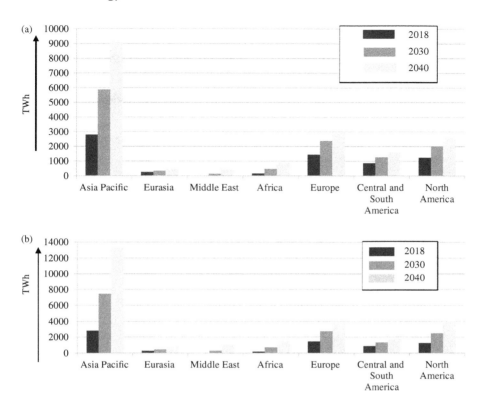

FIGURE 1.2 (a) Renewable electricity power by province and scenario, sustainable development, 2018–2040 in TWh (Gielen et al. 2019). (b) Renewable electricity generation by region and scenario sustainable development, 2018–2040 in TWh Sustainable Development Scenario (Gielen et al. 2019).

class 1 and class 2 locations. As a result, these sites have greater power-generating capacity and are thus more cost-effective. The installed power offshore wind by province and scenario (2018–2040) is shown in Figure 1.3.

INDIA's wind power capacity has significantly improved, and it is economically feasible to build class 3 sites. The current capacity of 65,000 MW was calculated using historical data. The future demand would be several times more than it was previously if class 3 sites were included. While exact estimates are yet to be published, experts believe the capacity is in excess of 150,000 MW. This increase in opportunity would ensure the industry's explosive growth for many years to come.

1.2.2 Wind Power Is Expected to Reach 43 GW by 2030

In view of the aforesaid facts – higher future potential and lower costs – the wind sector will continue to grow at a rapid pace. India added 1.2 GW of wind power in 2009, and more than 2 GW in 2010. If the current pattern continues, India's annual capacity additions could reach 4 GW by 2030, bringing total capacity to over 43 GW by 2030. The estimate is on the low end of the scale.

TABLE 1.1
Annual Average Renewable Asset by Scenario (billion USD 2018) (Dorraj et al. 2021)

	2018	Stated Policies 2019–2030	Stated Policies 2031–2040	Sustainable Development 2019–2030	Sustainable Development 2031–2040	2018 is the Year of Change vs 2031–2040	
Power generation from renewable resources	304	329	378	528	636	24%	109%
Wind	89	111	122	180	223	37%	151%
PV (solar power)	135	116	125	179	191	−7%	41%
End-use industries	25	117	139	124	145	456%	480%
Total	329	456	517	652	781	57%	137%
Cumulative		5477	5166	7829	7802		

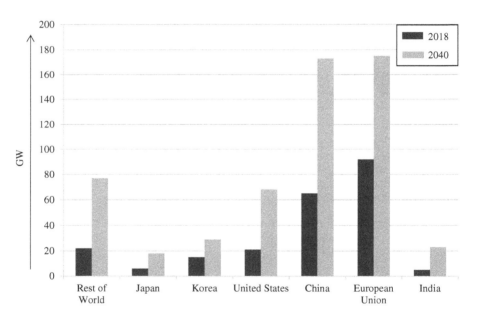

FIGURE 1.3 Installed power offshore wind by province and scenario, 2018–2040 in GW Stated Policies Scenario (Gielen et al. 2019).

1.2.3 Future Prospects for Solar

India's National Solar Mission aims to achieve a potential of 20,000 MW by 2022. The Indian government's strong market-based mechanisms in solar field offer captivating opportunities to developers and investors into the country, which directly results in ongoing cost reductions. The government's strategy in the solar field may promisingly propel solar energy to new heights in the future.

1.3 RENEWABLE ENERGY TECHNOLOGIES AND ONGOING DEVELOPMENTS

The pollution level throughout the world is now facing an alarming situation as the WHO indicated that about seven million people (per year) are dying from air pollution itself. Consumption of fossil fuels for electric power generation and transportation requirements is the main reason for such alarming pollution level situation. Now it doesn't matter that fossil fuels are in abundance or not as it is becoming somewhat impossible to lift the burden of using fossil fuels for our needs. The whole world is now looking for alternative sources of energy for electricity and transportation requirements which should be both sustainable and environmentally clean. Because of this, the focus of scientists is constantly shifting toward power generation technologies based on renewable energy. Renewable energy resources serve as an environmentally clean option for generating electricity, transportation requirements (electric vehicles), and renewable heating and cooling applications. In the succeeding sections, the renewable energy technologies like wind, solar, biomass, ocean energy systems and geothermal energy systems are discussed.

1.4 WIND ENERGY

Before the 20th century, wind energy was used mainly for water pumping (for salt making), sailing ships, grinding grains, etc. The first wind turbine for producing electricity (Nikitas et al. 2019) was built in 1887 at Cleveland, Ohio by Charles F. Brush (shown in Figure 1.4). Since then, by the mid-20th century, wind energy technology for electricity generation started getting used worldwide (Belyakov 2019) as a viable and capable technology. Starting from sailing of ships to generation of electricity, wind energy still remains one of the cleanest energy sources to lessen carbon foot prints or to reduce fossil fuel dependency. The basic fuel for this generation of electricity is flow of wind which has its advantage of being free and nonpolluting, but with the same, it has a limitation of its intermittency.

Various advantages and limitations of wind energy utilization are shown in Table 1.2. The succeeding subsections will discuss the basic principle, technical features, and economic and environmental analyses of wind power generation.

1.4.1 BASIC PRINCIPLE

The basic principle includes the conversion of wind's kinetic energy into electrical energy. The energy of flowing wind turns the blades of the wind turbine (producing rotational kinetic energy) which is connected to the gearbox through a horizontal shaft. The gearbox increases the speed of rotation and converts this rotational kinetic energy to electrical energy using generators. Such wind turbines are called horizontal-axis wind turbines (HAWT). There are also vertical-axis wind turbines (VAWT), but these are not used as much as HAWT (Belyakov 2019). There are factors like speed of wind and rotor blade's size, which decide the energy generated by wind (Yao). Figure 1.5 shows the energy conversion in wind power system.

FIGURE 1.4 The first automatically operated wind turbine with 18 m of height was built by Charles F. Brush in 1887 at Cleveland. It powered a 12 kW generator (Nikitas et al. 2019).

TABLE 1.2
The Benefits and Limitations of Wind Energy

Benefits	Limitations
Lowest greenhouse gas emission as compared to other technologies (Jacobson 2009).	Discontinuous flow of wind leads to intermittent electricity supply.
It is a fully sustainable energy source.	Quality sites are remotely located and requires transmission lines.
It occupies less space and there can be situated together with farming.	Scarcity of neodymium (Nd), used in magnets of wind turbines (Pavel et al. 2017).
Wind farms consumes less water as compared to other plants like thermal and biomass (Letcher 2017).	Wildlife impacts: birds and bats fatality (Willis et al. 2009 and Johnson 2005).
Create jobs and necessary in economic growth.	Environmental impacts: Noise Pollution and aesthetic pollution (Kaldellis et al. 2012).
Zero fuel cost with very less operation and maintenance cost. Also, construction time is less.	Difficult and pricey to repair the rotating parts which are placed high off the ground.

1.4.2 Technical Aspects of Wind Turbines

The main components of wind turbines include rotor, gearbox, generator, tower and its foundation, etc. The working principle to produce electricity from wind energy requires all these components of wind turbines. Each of these main components is explained briefly in the next subsections.

1.4.2.1 Rotor

Rotor receives the kinetic energy of wind using its rotor blades and converts it into rotational kinetic energy which is provided to the rotor shaft with the help of rotor

Renewable Energy Scenario of the World

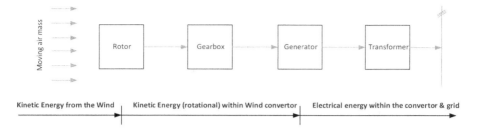

FIGURE 1.5 Energy conversion in wind power system.

hub, and this rotor shaft is connected to the gearbox. The size and shape of rotor blades mainly depend upon the plant-installed capacity (Kaltschmitt et al. 2007). The material used to make the blades usually are glass fiber-reinforced plastics. The main criteria for selecting the material is the manufacturing cost depending upon the requirements of rotor's strength of breaking, specific weight, elasticity, strength, etc. The connection between rotor shaft and rotor blades is done using a rotor hub (high quality cast iron) which are of three types: rigid or hinge less hub, teetered hub, and flap hub (Kaltschmitt et al. 2007).

1.4.2.2 Nacelle

It basically houses the tower-top components including gearbox with main shaft, generator, brakes, cooling system, etc. (Rao 2019). It protects the components from weather and also allows the rotor to face the wind if the direction of wind changes. The nacelle changes its orientation about vertical axis with the change in direction of wind by using a yaw motor equipped with it (Rao 2019). Yaw mechanism rotates the nacelle to bring the rotor blades to face the wind especially when the wind changes its direction.

1.4.2.3 Gearbox

The rotational speed of rotor is very small, and gearbox helps to increase this rotational speed to match it with the generator used. The gearbox is required to increase this rotor speed of wind turbine to about 75–150 times to match the generator speed (3,000 rpm and 1,500 rpm for 2 poles and 4 poles conventional generator at 50 Hz). The difficulties related to the gearbox arise with the surviving from loads, vibrations, heating (through friction), and shocks applied on it. Therefore, it is termed as a high-maintenance unit and requires high repair service in about every 5 years.

1.4.2.4 Generator

The wind turbine generator takes the mechanical energy from the rotor of wind system through gearbox and then coverts it to electrical power. The generator can be synchronous and nonsynchronous (induction generator) types. The most commonly used generators for wind turbines are double-fed induction generator (DFIG), squirrel cage induction generator (SCIG), and permanent magnet synchronous generator (PMSG) (Beainy et al. 2016). The advantages and disadvantages of each type of generator used are given in Table 1.3 (Beainy et al. 2016).

TABLE 1.3

Advantages and Disadvantages of Different Generators Used in Wind Turbines

	DFIG	SCIG	PMSG
Advantages	Rugged, brushless, lightest, low cost, wide range of speed, high efficiency and integration with grid is smooth	Simple and robust, operation requires no brushes, escapes the grid's short circuit power	No gearbox is required and therefore low maintenance cost, more reliable and longevity
Disadvantages	Difficulty in dealing with ride-through of grid fault, bearing and grid faults leads to less reliability	For its operation, it needs convertors (two full scale), no gearless operation capability	Expensive, twice the outer diameter of gearless PMSG as compared to conventional SCIG

1.4.2.5 Tower and Foundation of Wind Turbine

The height of the tower is not fixed by any law but the fact that it should be twice or thrice the length of the rotor blades. Also, the kinetic energy of wind increases with height and so as electric power. Therefore, the height of the tower is also based upon the factors that include power generation vs. cost considerations. Also, on the other hand, the whole structure of wind turbine system with its tower requires a foundation. Turbine foundation is a very important component for making the huge structure of the wind turbine operate without falling. The different types of foundations include gravity foundation, monopile foundation, and tripod foundation (Kaltschmitt et al. 2007).

1.4.3 Control Technique Requirements

To control the output power from the wind turbines, one can either adjust the blade angle (facing wind) of the rotor or completely rotate the wind turbine. These methods are discussed below.

1.4.3.1 Pitch Adjustment Control Method

It is a very effective method to limit the output power by varying the pitch angles of the rotor blades so as to vary the aerodynamic force of high wind speed. Stalling and furling methods are used to change blade pitch angle (National Instruments 2020).

1.4.3.2 Yaw Mechanism

In this method, the entire wind turbine rotates (horizontally) so as to make the wind flow fall perpendicular on the surface area of the rotating blades. This maximizes the output power and prevents the losses which can occur because of misalignment of turbine with the oncoming wind (National Instruments 2020).

Renewable Energy Scenario of the World

1.4.4 Economical Analysis and Environmental Impacts

1.4.4.1 Economic Analysis

The various economic aspects of wind energy system are given in Figure 1.6. The economic components of wind energy system include availability of turbine (during maintenance and repairs), economic lifetime, energy efficiency (low efficiency leads to lesser investment returns), wind regime (wind speed distribution throughout the year required for wind turbine), investment costs (capital cost and financial cost), and recurring cost (operation and maintenance costs). To harness useful energy from wind energy system, it is necessary to consider all these economic aspects to understand the cost-effectiveness of the system (Yao et al. 2011).

1.4.4.2 Environmental Impacts

Although the wind energy system is considered to be an environmentally clean source of energy, it still has many negative impacts on the environment. These negative impacts include fatality of birds and bats, noise pollution, emissions of greenhouse gases (mainly due to production of concrete and steel required for foundation of turbine), and land surface temperature change (Gupta et al. 2021)

1.5 SOLAR ENERGY

The solar energy has always been an important part of human life. In its early ages, the solar energy was used for producing fire (using magnifying glasses), lightening a room using mirrors, and cooking purposes. But now, the solar energy is used to produce electricity directly using solar PVs as concentrated solar power systems (CSP) (Kabir et al. 2017). Other than that, one also cannot neglect the fact that solar energy is the basic source of energy for other green technologies like wind, wave, tidal, and bioenergy.

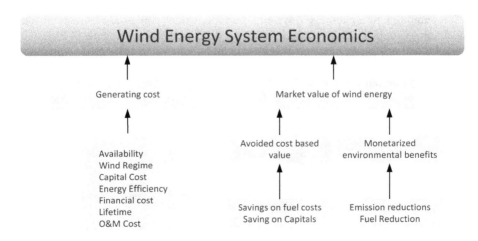

FIGURE 1.6 Wind energy system economical components (Yao et al. 2011).

1.5.1 Solar Power Technology

In this section, the two viable technologies used for extracting solar energy from sun and converting it into electricity or other heating applications are discussed. These two solar technologies are PV and solar thermal CSP technologies. These technologies give existence to the fact that solar energy can be used to generate electricity and make it even more popular. Solar thermal (CSP) technology collects solar energy and converts it into heat, which is further converted to electricity, while PV technology directly converts solar energy to electrical energy using semiconductor devices. There are other solar thermal technologies (nongrid) which include water-heating systems, solar dryers, solar distillations, and solar cookers. Figure 1.7 shows the SolDry (solar drying) systems designed and developed by the National Institute of Solar Energy (NISE) which benefits the farmers of Ladakh (NISE 2020).

1.5.2 CSP Systems

As discussed earlier, both CSP and PV technologies use solar energy for their operation. A CSP plant (Awan et al. 2019) (shown in Figure 1.8) consists of three main parts which include solar receiver system, thermal energy storage (TES) system, and power block. The solar receiver system concentrates light energy coming from the sun onto the receiver and converts it into heat. This heat is absorbed by heat transfer fluid (molten salt or synthetic oil) which transfers the collected heat of the sun to the power block. This thermal energy at power block gets converted into electricity using steam turbine generators. The excess thermal energy from the sun gets stored in TES system. It uses molten salt to absorb the solar heat (from the solar receiver), which is stored in a hot tank. TES system is used during the absence of the sun and avoids the shutdown of CSP plant (Awan et al. 2019). There are three types of solar concentrating systems (Figure 1.8) which are solar tower, parabolic trough, and parabolic dish (Yao et al. 2011).

1.5.3 PV Systems

Compared to CSP systems, PV systems are more extensively used commercial solar technology for producing electricity directly. It is based on the principle of PV effect. A solar cell or PV cell made up of semi-conducting materials is the basic component of the PV system. P-N junction is formed by doping the solar cell, and this structure acts as an internal electric field. The electron-hole pairs are formed due to the photons hitting (when sunlight falls on the cell) on the surface of the cell. Due to the

FIGURE 1.7 SolDry systems distributed among Ladakh farmers (NISE 2020).

Renewable Energy Scenario of the World

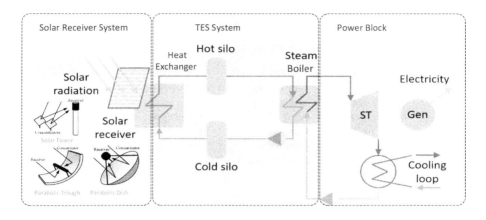

FIGURE 1.8 Layout of CSP plant.

internal electric field, the electrons and holes move toward the respective positive and negative electrodes. Hence, the current flows through the load connected between the electrodes. The output voltage of PV cell is very low (0.5–0.6 V), and therefore, they are connected in series to increase to form a solar module. The desired voltage-current output is obtained from the series and parallel connections of these solar modules. The semiconductor material is used for manufacturing PV cells based on silk-screen process building (Kaltschmitt et al. 2007). Based on the type of semiconductor material, the PV cell is named accordingly. As shown in Figure 1.9, there are three generations of PV cell technology (Belyakov 2019). There are two types of PV systems which include grid-connected PV systems and stand-alone PV systems (Prasad and Bansal 2011). Grid-connected PV systems use the net-metering method where the net meter runs in both directions which also calculates the energy produced by customers. While in the stand-alone system, it is necessary to match the PV system with the system load (Prasad and Bansal 2011).

Both the technologies (CSP and PV) have their respective advantages and disadvantages. Table 1.4 compares these two technologies of solar energy market with respect to the technology used, storage capability, efficiency, and market capture.

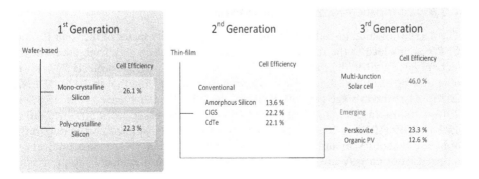

FIGURE 1.9 PV cell technologies.

TABLE 1.4

Comparison Between CSP and PV Systems

	Technology	Energy Storage	Efficient	Market
CSP System	It converts the concentrated sun light energy to heat energy and then to electrical energy using steam generators.	Capable of storing thermal energy through TES systems. Therefore, it can be used even when there is no sun light.	It collects the solar energy more efficiently as compared to PV system.	Energy price is higher (compared to PV systems) therefore with respect of PV systems, growth of CSP systems is not that high.
PV System	It directly converts the solar energy to electricity using PV cells.	Not capable to store energy in the thermal form and storing electricity energy (in batteries) is not easy.	It is less efficient to collect the solar energy (compared to CSP systems) required for the operation of PV cell.	Economically more feasible and therefore its growth rate is higher than that of CSP systems.

1.5.4 ENVIRONMENTAL IMPACTS

Solar energy is one of the cleanest energy sources, but still, the electricity generation from solar energy has some negative impacts on the environment. These negative impacts include strain on clean water availability especially in the arid areas over the world where high intensity of sunlight is available. This clean water is used as condenser cooling water. In some cases, the high bird mortality at the solar power plant has also been found (Holbert 2011). The PV system uses heavy metals like cadmium telluride for manufacturing of PV cells which are toxic for the environment. Following that, the other harmful impacts on the environment (because of solar energy) are land usage, ecological impacts, chemical spills, and recycling problem of solar panels (Holbert 2011).

1.5.5 ECONOMIC ANALYSIS OF SOLAR ENERGY

The decreasing total installed cost and increasing capacity factor of the solar PV and CSP systems lead to the reduction in levelized cost of electricity from the solar PV technology. In 2019, the total average installed cost of solar PV projects 79% lower than that in 2009, and that of CSP system is 36% lower than that in 2009 (IRENA 2020). Also, the average capacity factor of the solar PV is increased from 13.8% in 2010 to 18.0% in 2019, and that of CSP systems is increased from 30% in 2010 to 45% in 2019 (IRENA 2020). The reduction of the total installed cost is due to the reducing cost of PV module over the period of time. The increased mass production, reduced labor costs, increased efficiency, and optimized manufacturing process of PV modules are the main reasons behind the drop in the average installed cost of solar PV systems. The main factor behind the reduced installed cost of CSP systems

Renewable Energy Scenario of the World

is the reduced cost of electricity storage with the CSP systems. The increased capacity factor of the solar PV and CSP systems is mainly achieved by selecting the better locations with high irradiation, more usage of improved tracking devices, improved electricity storage capacity, and reduction in losses. The continuous decline in the cost of solar PV and CSP systems makes the solar energy usage more viable compared to other renewable technology

1.6 BIOMASS ENERGY

Biomass energy (or bioenergy) is a unique type of renewable energy source which is basically a stored form of sun's energy present in organic materials (biomass) like plant, trees, crops, animal wastes, etc. Biomass is an organic material which is renewable and derived from living (or recently dead) biological organisms like plants and animals. It is renewable in nature because it basically comes from sun and can be regrown in a short period of time. Bioenergy uses biomass materials to undergo various conversion procedures (thermal and biochemical) to get other forms of useful energy like heat and electrical. It also produces some solid, liquid, and gaseous fuels from biomass using the same conversion procedures.

Similar to biomass, fossil fuels also come from organic matter, and they both produce carbon dioxide when converted into electricity and heat. But the main differences between biomass and fossil fuels are hidden behind the following facts:

- Biomass is renewable in nature, while fossil fuels are nonrenewable.
- Biomass is carbon-neutral, i.e., it takes the carbon dioxide from the environment (photosynthesis) and gives it back during the conversion procedures for producing electricity and heat. Therefore, it maintains the balance in carbon cycle.
- Most biomass feedstocks have lower bulk densities than fossil fuels except liquid biofuels which somewhat have comparable bulk densities.
- The fossil fuels and biomass both originate from organic materials, but fossil fuels are formed by decomposition of organic matter for millions of years. Compared to that, biomass is derived from recently living organic materials.
- Biofuels like ethanol and biodiesel (liquid fuel from biomass) have nontoxic effects (low sulfur and ash content) compared to the use of fossil fuels.

1.6.1 BIOMASS: TYPES AND HARVESTING

Biomass refers to the organic matters coming from plants and animals which are recently living and have stored sun's energy. The classification into different types of biomass (Seveda et al. 2011) is based on their sources. The following are the different types of biomass:

- Biomass from Waste: Animal and Municipal Waste

  It includes farm slurries and poultry litters from animal farming (cattle and pig farming). Other animal waste from slaughterhouses and fish processing is also a good source of biomass. While municipal waste includes commercial and residential wastes like human excreta, food, paper, etc., it also includes sewage wastes in liquid form.

- Agricultural Biomass
  This includes agricultural residues (like stalks, leaves, branches, pruning's waste, etc.) and by-products of agricultural processes.
- Forest Biomass
  Forest trees are the main contributor with its parts like trunks, leaves, barks, and roots. The by-products (wood chips and saw dust) from the wood industry processes are also used as biomass.
- Energy Farming
  The term energy farming is used on a broader way for production of biomass feedstocks in a short period of time. It includes some certain types of crops, trees, and shrubs, which require relatively less harvesting time and are used as a biomass feedstock.
- Marine Biomass
  It includes algae and other marine biomass (like kelps, water hyacinths, etc.) which are also the source for producing biomass feedstock.

There are several harvesting methods which make the plants renewed by sprouting. These methods are coppicing, pollarding, lopping, pruning, and thinning method (Seveda et al. 2011).

1.6.2 BIOENERGY CONVERSION TECHNOLOGIES

The various biomass conversion technologies are used to produce secondary energy sources (heat, electricity, biofuels) by converting biomass feedstocks (lignocellulose, microalgae, animal and food waste, etc.). These conversion technologies are divided into four categories namely thermal, chemical, thermochemical, and biochemical conversions (as shown in Figure 1.10).

In the direct combustion method, burning of solid biomass takes place just like it is done for other conventional fossil fuels (Belyakov 2019). This technology is used for producing heat and electricity. The thermochemical biomass conversion technology uses the following three methods (Jahirul et al. 2012):

- **Pyrolysis:** This process includes heating of solid biomass for producing biofuels (bio-char and bio-oil). The heating is done in the absence of oxygen at a temperature of 350°C–800°C and beyond.
- **Torrefaction:** It is similar to pyrolysis but performed at lower temperature to obtain fuels (bio-coal and charcoal) with improved properties.
- **Gasification:** This process gives synthesis gas (gaseous biofuels) by making the solid biomass subjected to face hot steam and air.

Another conversion technique, biochemical biomass conversion, includes the process of fermentation and anaerobic reaction. Fermentation is a process of converting sugar (coming from energy crop) to ethanol. In this process, yeast is mixed with biomass, and under specific conditions, this mixture is then allowed to ferment, while anaerobic digestion requires degradation of biomass (food waste, slurries and crop residues)

Renewable Energy Scenario of the World

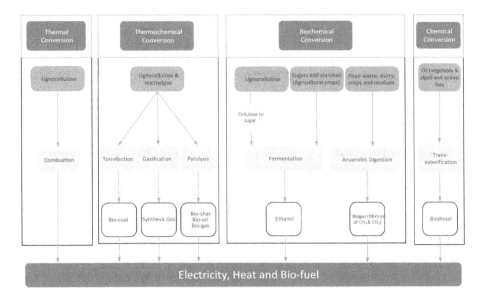

FIGURE 1.10 Biomass conversion processes.

in the absence of oxygen with low heating. The primary fuel from anaerobic digestion is a mixture of methane and carbon dioxide (Belyakov 2019) which is further used for producing electricity and heat applications, whereas biodiesel is produced by transesterification (chemical conversion) of vegetable and algal oils (Seveda et al. 2011).

1.6.3 Environmental Impacts of Biomass Energy

There is no doubt that using biomass as an alternative for fossil fuels is environmentally more beneficial. But the uncertainty remains in the fact that how much beneficial it is to the environment. Practically it is not that simple to consider biomass as carbon-neutral. There are several negative environmental impacts of using biomass as an energy source which are related to water quantity and quality, greenhouse gas emissions, deforestation, biodiversity, and soil erosion (Wu et al. 2018). The scarcity of water quantity comes with the cultivation of bioenergy crops like corn used for corn ethanol production (biofuel). The quality of water is also a concern with the cultivation of bioenergy crops as it increases the risk of high concentration of nitrate in the water sources. Although biomass is considered as carbon-neutral, still, biomass use for energy production leads to emission of NO_2 which is further responsible for global warming. Therefore, it is essential to control the emission of NO_2 from biomass usage. Conversion of land required for cultivating the bioenergy crops is one of the main factors for impacting the biodiversity. Soil erosion is also very common which is due to land use change and expansion of corn acreage. Other than carbon dioxide, burning of wood biomass emits additional harmful gases like nitrogen oxides, sulfur dioxide, and carbon monoxide. So, while concluding the fact that biomass is environmentally clean, one should rather consider it as a better alternative to fossil fuels.

1.6.4 Economic Analysis of Biomass

Unlike solar and wind energy technology, the biomass energy for electricity production requires feedstocks to be produced, transported, and stored (IRENA 2012). About 40% of the cost required for the production of electricity from biomass is dependent on its feedstock production and management. The cost of power generation from biomass is wide-ranging depending upon the different energy technologies. The capital expenditure cost is dependent on the factors like components of power plants, feedstock management and production, technology included, plant size, construction, and other engineering aspects (IRENA 2012; Carneiro and Ferreira 2012). The operation and maintenance costs are both fixed and variable. The fixed operation and maintenance costs include labor cost, scheduled replacement and services of equipment, and aspects like insurance, etc., while the variable operation and maintenance costs are dependent on unscheduled maintenance, disposal of ash, and transformational costs (IRENA 2012; Carneiro and Ferreira 2012). The biomass energy technologies still have a huge scope of its expansion to contribute more as a cleaner and more efficient renewable source of energy. It requires better policies, laws, permits, and political support to draw the attention of the private sector to invest more on biomass energy technology.

1.7 OCEAN ENERGY

Covering about 70% of the Earth's surface and holding 97% of water of our planet, oceans can't get neglected for considering them as an enormous source of energy. The energy technologies used to harness electric energy from oceans include tidal energy, wave energy, ocean current energy, and ocean thermal energy. All these energy technologies are renewable in nature. These energy technologies are not commercially available yet and still under research and development stage. The reason for not being commercially available for producing electricity is related with its high capital cost (MNRE 2021). Even after having a huge potential, the ocean energy technologies still lack in their global energy contribution. It requires more research and development to make them available commercially widely. In the following subsection, the different ocean energy technologies with their impacts on the environment are discussed.

1.7.1 Different Types of Ocean Thermal technologies

There are mainly four types of ocean energy technologies which include wave energy, tidal energy, ocean current energy, and ocean thermal energy. The block diagram of these ocean energy technologies is shown in Figure 1.11. The mechanical forces and thermal gradient related to the ocean results in the types of ocean energy technologies one should use. The mechanical forces are due to the wind flowing over the ocean and also due to the gravitational pull of moon. The latter creates tides and currents, while the flow of wind over the ocean surface creates waves. Thermal radiations of sun make the surface of the ocean hotter, while depths remain colder. This thermal difference within the layers of oceans can be utilized to create electrical power.

Renewable Energy Scenario of the World

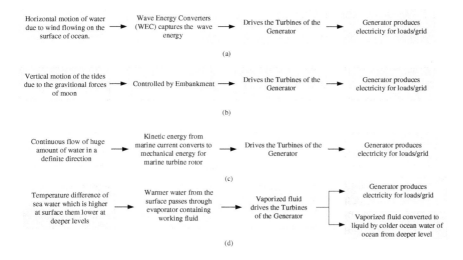

FIGURE 1.11 Block diagram of various ocean energy technologies. (a) Wave energy. (b) Tidal energy. (c) Marine current energy. (d) Ocean thermal energy.

Wave energy converters (WEC) are used to convert the energy of ocean waves into electricity. The waves in the oceans are produced by the wind flowing on their surface, thereby making it as a concentrated form of solar energy. The different types of WEC are attenuators, point absorbers, overtopping terminators, oscillating wave surge converter, submerged pressure differential, and oscillating water column (Dolores et al. 2017). These WEC devices capture the mechanical energy from the up-down movements of waves and convert it into electrical energy. Another way to harness electrical energy from ocean is to make use of tides (low and high) in the ocean similar to that of a dam. There are around 20–40 locations throughout the world which can be used to as a potential harness electrical energy from these low and high tides of the oceans (U.S. Department of Energy Office 2009). While considering the ocean currents, the factors responsible for the formation of ocean currents are winds, tides, changes in water densities, and rotation of the Earth. Harnessing electrical energy from ocean current energy is similar to that used by wind turbines for wind energy systems. The turbines used are axial-flow or cross-flow situated turbines. The axial-flow turbines are further categorized with shrouded rotor or open rotors (U.S. Department of Energy Office 2009). Another method for converting the energy from ocean comes from the fact that there is a temperature difference between the surface and lower layers of the oceans. The technologies used for generating electrical energy related to this fact are open loop and closed loop ocean thermal energy conversion (OTEC) technologies (U.S. Department of Energy Office 2009). All these ocean energy technologies are still not in its mature stage as compared to other renewable technologies. But one cannot ignore the fact that oceans are enormous source of energy, and it leaves with a huge scope for its future research and developments of its technologies.

1.7.2 Environmental Impacts

There is a huge significance of oceans for producing oxygen (around 50%) and absorbing carbon from the atmosphere. However, at the same time, there is a risk of polluting ocean from the new technologies of developing the electrical energy from the ocean energy. So, it becomes really important for estimating the environmental impacts of these new ocean technologies required for generating electrical energy from oceans. The following are the related environmental impacts of harnessing electricity from ocean energy technologies:

- Variations in hydrodynamics due to the dykes used in tidal energy conversion influences the local population of fishes, seabirds, and mammals (Haverson et al. 2018).
- The species near the vicinity of the OTEC plants experience variations in size and quantity. This is mainly because of the deep-water discharges by the OTEC technologies.
- The constructions or installation leads to high level of noise which influences the presence of key species, their colonization, and population. Invasion of new species occurs which further disturbs the original marine fauna. The establishments under the oceans are changing the marine life around it (Rivera et al. 2020).
- The species experience variations in their migration routes as the installations of the ocean technologies limit the movements of the species (Rivera et al. 2020).
- It creates collision risks for sea mammals with the installed devices of the ocean technologies (Rivera et al. 2020).
- The support required for the installation of ocean technologies can lead to changes in seabed structure (Rivera et al. 2020).

1.7.3 Economic Analysis of Ocean Energy Technology

The ocean energy conversion to electricity is still at its precommercial state. In India, the ocean renewable energy technology is still considered as a new technology with high capital cost, and it is still in research and development stage, which is not yet used at commercial level (MNRE 2021). The capital cost of ocean renewable energy is dependent on the types of factors like marine devices, location, cables/pipelines, and installations. Since ocean energy is a relatively new technology, therefore, the capital cost is always subject to variations depending upon the technology, locations (depth of water and distance from shores), installation charges, and other factors (Dalton et al. 2015). The operating cost is mainly dependent on the maintenance which is both scheduled maintenance and unscheduled maintenance. The scheduled maintenance deals with all the service tasks required from time to time, while the unscheduled maintenance is dependent on the monitoring system of the ocean energy technology (Dalton et al. 2015). The ocean energy industry is a new and emerging sector, and therefore, its accurate economic analysis still requires some time for its complete commercial acceptance and development.

1.8 GEOTHERMAL ENERGY

Geothermal refers to producing energy from the internal heat of the Earth. The internal heat of the Earth is generated from radioactive decay of minerals and continual heat loss from the Earth's original formation. The Earth's crust is made from rocks and water. Deep below the Earth's surface, there is a layer of hot-molten rock called magma. The heat present at 10 kms below the Earth's surface contains 50,000 times more energy than all the oil and natural gas resources in the world. The use of this heat energy in a right way can be used for heating and cooling purposes or to create clean electricity. The geothermal energy is a renewable source of energy which is nonpolluting and consistently reliable. Figure 1.12 shows the year-wise (2010–2019) installed capacity of the geothermal energy (IRENA 2021). The worldwide installed capacity of geothermal energy (IRENA 2021) is about 14 GW (2019) which is not well explored as compared to its capability. Therefore, geothermal energy technology remains an important scope for its exploration and development.

1.8.1 Resources of Geothermal Energy

The resources of geothermal energy can be categorized in the following three categories (Haverson et al. 2018):

- Geothermal reservoir
  In this category, the water on the Earth's surface leaks down to the hot crust of the Earth and gets heated. Due to process of convection, this heated water rises up to the surface forming hot springs, geysers, fumaroles, or hot mud

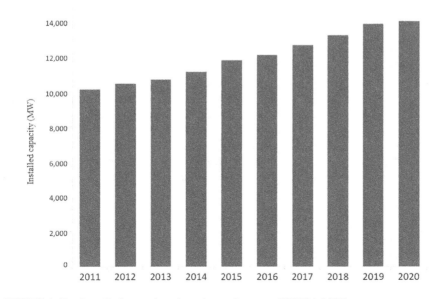

FIGURE 1.12 Installed capacity of geothermal energy (IRENA 2021).

holes. If the heated water gets blocked while coming upwards, then by drilling bores, a path can be made for it to come up at the surface (Breeze 2019).

- Hot dry rocks

  Due to some irregularities in the crust, the rock near the Earth's surface is much hotter than normally. This can be used as a source of geothermal energy under such conditions by drilling the surface and pumping the water (or any heat transfer fluid) out through it (Breeze 2019).

- Magma

  Magma is extremely hot-molten liquid or semi-liquid present under the Earth's surface. When it reaches the surface, it is called lava. It is also considered as a very rich source of geothermal energy but simultaneously it has a limitation of extracting this energy formed within the earth's outer crust (Breeze 2019).

1.8.2 Geothermal Energy Technologies

The different technologies for geothermal energy depend upon its direct and indirect usage. The direct use of geothermal energy includes heating and cooling of buildings (Breeze 2019). This heating and cooling application of geothermal energy makes technology like ground source heat pumps and deep enhanced geothermal systems (Breeze 2019). On the other hand, indirectly the geothermal energy is used to produce electricity. The production of electricity from geothermal energy uses technologies like dry steam power plants, flash point power plants, and binary cycle power plants (Breeze 2019). All these technologies are based on the type of geothermal resource location. The direct dry steam power plants require geothermal reservoirs which give high-temperature dry steam alone. The geothermal reservoirs with steam and liquid brine mixture require a flash point power plant for electricity production. The binary cycle power plants are used where the temperature of the geothermal location is relatively low.

1.8.2.1 Dry Steam Power Plants

The geothermal reservoirs with high-temperature dry steam are required for this type of geothermal power plant. This high-temperature steam is extracted by making boreholes under the ground of the geothermal locations, and filters are used with the boreholes to filter the rock pieces coming out with the steam. The steam turbines use this dry high-temperature ($180°C$–$350°C$) steam to get rotated and to produce electricity (Figure 1.13a). The steam used here also contains about 2%–10% of other gases like CO_2 and H_2S. So, the exhaust of steam after passing through the steam turbine is required to be condensed and treated for pollutants as well (Breeze 2019). Sometimes for economic reasons, the geothermal fluid is released in the atmosphere instead of injecting it back.

1.8.2.2 Flash Point Power Plants

The geothermal reservoir fluid is seldomly in the dry steam form, and usually it is a mixture of steam and liquid brine. This fluid having a high pressure is passed through a vessel with low pressure inside. A major part of the liquid flashes into steam because of this sudden change in the pressure which increases the percentage of the steam in the fluid. This steam is separated and fed to the steam turbine to produce electricity (Figure 1.13b) just like in dry steam power plants.

Renewable Energy Scenario of the World

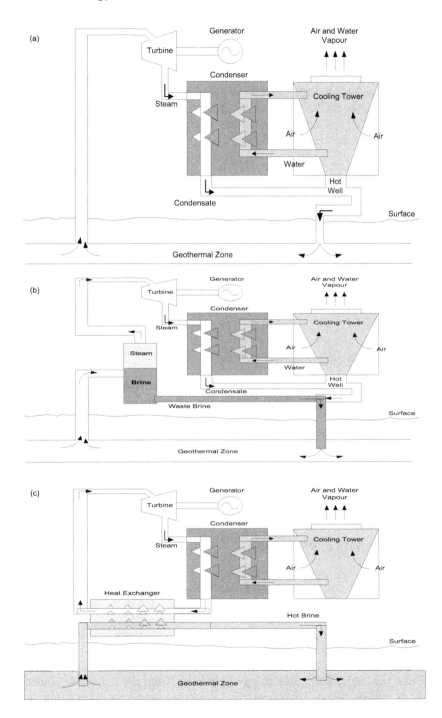

FIGURE 1.13 (a) Dry steam power plant. (b) Flash point power plant. (c) Binary cycle power plant.

1.8.2.3 Binary Cycle Power Plant

The dry steam and flash point plants are not very efficient when the temperature of the geothermal fluid is low, i.e., between 100°C and 170°C. For such a case, the binary cycle power plant is used to extract the energy out of it to produce electricity. As shown in Figure 1.13c, the binary cycle plant uses the resource geothermal fluid to heat the process fluid (ammonia and water mixtures) through heat exchanger (IRENA 2017). The boiling and condensate points of the process fluid are according to the geothermal temperature available.

1.8.3 Cost Analysis

Extracting useful energy from geothermal resources requires high capital cost for big constructions but lower operating and fuel cost. The cost analysis depends mainly on three factors that include exploration of geothermal locations, construction of steam field, and cost of power plant (Breeze 2019). The expenditure includes several million dollars for finding appropriate geothermal reservoirs, and sometimes it may not be successful. The cost of construction of steam field depends on the size of plant and the number of boreholes required to be drilled. The plant construction cost is dependent on the factors like geothermal location and temperature (Breeze 2019; IRENA 2012).

1.8.4 Environmental Impacts

The environmental impacts of the geothermal energy utilization are much lower as compared to the use of fossil fuels. It is mainly due to the use of high-temperature geothermal systems for energy conversion. Table 1.5 shows the possible environmental impacts of the development of the low and high geothermal systems. These environmental impacts are due to the various reasons related to the drilling operations, mass withdrawn of geothermal fluids, and liquid and gaseous waste disposals (Hunt 2000). All these factors related to the geothermal systems lead to the following effects like destruction of forests, noise pollution, ground water contamination, degradation of geothermal features, ground water depletion, and induced seismicity (Hunt 2000).

1.9 CONCLUDING REMARKS

The renewable energy technologies are the need of the future energy requirements and have been discussed accordingly. This need is not only because of the depletion of fossil fuels but also due to the alarming level of pollution in the environment. The time has already ringed the bell to switch the use of fossil fuels requirements for electricity, transportation, and other uses. That is why renewable energy technologies are needed because they are not only sustainable but also environmentally much friendlier as compared to the technologies based on fossil fuels. It has also been shown that the future pattern of energy harnessing from the renewable sources is growing at very high rate. This chapter has presented these renewable energy technologies considering its present and future scenarios, working operations, environmental impacts, and cost analysis.

TABLE 1.5
The Possible Environmental Impacts of Geothermal Power Plants (Rivera et al. 2020)

		Drilling Operations	Liquid Waste Disposal		Gas Waste Disposal			Mass Withdrawal		
			Impacts on Living Organism							
	Noise	Contamination of Ground Water with Drilling Fluid	Surface Discharge	Reinjection	Impacts on Living Organisms	Microclimate Influences	Degradation of Thermal Features	Depletion of Ground Water	Ground Temperature Change	Mass Withdrawal
Low Temperature System	Slight Effect	Medium Effect	Slight Effect	Slight Effect	No Effect	No Effect	No Effect	Slight Effect	No Effect	No Effect
High-Temperature Vapor Dominated	Medium Effect	Medium Effect	Medium Effect	Slight Effect	No Effect	Slight Effect	Slight Effect	Medium Effect	Slight Effect	Slight Effect
Water Dominated System	Slight Effect	Medium Effect	Medium Effect	High Effect	No Effect	Medium Effect	Slight Effect	High Effect	Medium Effect	Medium Effect

REFERENCES

Abbaszadeh, P., Maleki, A., Alipour, M., and Maman, Y.K. 2013. Iran's oil development scenarios by 2025. *Energy Policy*, Volume 56, 2013, Pages 612–622. ISSN 0301-4215.

Awan, A.B., Zubair, M., Praveen, R.P., and Bhatti, A.R. 2019. Design and comparative analysis of photovoltaic and parabolic trough based CSP plants. *Solar Energy*, Volume 183, 2019, Pages 551–565. ISSN 0038-092X. https://doi.org/10.1016/j.solener.2019.03.037.

Beainy, A., Maatouk, C., Moubayed, N., and Kaddah, F. 2016. Comparison of different types of generator for wind energy conversion system topologies. *2016 3rd International Conference on Renewable Energies for Developing Countries (REDEC)*, Zouk Mosbeh, 2016, pp. 1–6, doi:10.1109/REDEC.2016.7577535.

Belyakov, N. 2019. Chapter 16 - Wind energy. Editor(s): Nikolay Belyakov, *Sustainable Power Generation*, Academic Press, Cambridge, MA, 2019, Pages 397–415, ISBN 9780128170120.

Breeze, P. 2019. Chapter 12- Geothermal power, Editor(s): Paul Breeze, *Power Generation Technologies* (Third Edition) Newnes, London, 2019, Pages 275–291. ISBN 9780081026311. doi:10.1016/B978-0-08-102631-1.00012-2.

Carneiro, P., and Ferreira, P. 2012. The economic, environmental and strategic value of biomass. *Renewable Energy*, Volume 44, 2012, Pages 17–22, ISSN 0960-1481, doi:10.1016/j.renene.2011.12.020.

Dalton, G., Allan, G., Beaumont, N., et al. 2015. Economic and socio-economic assessment methods for ocean renewable energy: Public and private perspectives. *Renewable and Sustainable Energy Reviews*, Volume 45, 2015, Pages 850–878, ISSN 1364-0321, doi:10.1016/j.rser.2015.01.068.

Dorraj, M., Morgan, K., Alhajji, A., et al. 2021 *Global Impact of Unconventional Energy Resources*. Lexington Books, Lanham, MD. ISBN: 978-1498566094.

Dolores, E.M., José, S.L.G., and Vicente, N. 2017. Classification of wave energy converters. *Recent Advances in Petrochemical Science*, Volume 2(4), Page 555593. doi:10.19080/RAPSCI.2017.02.555593.

Gielen, D., Boshell, F., Saygin, D., Bazilian, M.D., Wagner, N., and Gorini, R. 2019. The role of renewable energy in the global energy transformation. *Energy Strategy Reviews*, Volume 24, 2019, Pages 38–50. ISSN 2211-467X. doi:10.1016/j.esr.2019.01.006.

Gupta, N., Joshal, K.S., and Tomar, A. 2021. Chapter 9- Environmental and technoeconomic aspects of distributed generation. Editor(s): Anuradha Tomar, Ritu Kandari, *Advances in Smart Grid Power System*, Academic Press, Cambridge, MA, Pages 237–263. ISBN 9780128243374. doi:10.1016/B978-0-12-824337-4.00009-6.

Haverson, D., Bacon, J., Smith, H.C.M., and Venugopal, V.X.Q. 2018 Modelling the hydrodynamic and morphological impacts of tidal stream development in Ramsey sound. *Renewable Energy*, Volume 126, Pages 876–887.

Holbert, K. E. 2011. Chapter 10 - solar thermal electric power plants. *Handbook of Renewable Energy Technology*. Copyright © 2011 by World Scientific Publishing Co. Pte. Ltd. Pages 225–243, ISBN-13 978-981-4289-06-1.

Hunt, T. M. 2000. Five lectures on environmental effects of geothermal utilization. Geothermal training programme. The United Nations University, Reports 2000. Number 1, 9–22.

IRENA. 2012. Biomass for Power Generation, Renewable Energy Technologies: Cost Analysis Series. Volume 1: Power Sector, Issue 1/5.

IRENA. 2017. Geothermal Power: Technology Brief, International Renewable Energy Agency, Abu Dhabi.

IRENA. 2020. Renewable Power Generation Costs in 2019. ISBN 978-92-9260-244-4, International Renewable Energy Agency, Abu Dhabi.

IRENA. 2021. https://www.irena.org/geothermal.

Jacobson, M.Z. 2009. Review of solutions to global warming, air pollution, and energy security. *Energy & Environmental Science*, Volume 2, Pages 148–173. doi:10.1039/B809990C.

Jahirul, M.I., Rasul, M.G., Chowdhury, A.A., and Ashwath, N. 2012. Biofuels production through biomass pyrolysis —A technological review. *Energies*, Volume 5, Pages 4952–5001. doi:10.3390/en5124952, energies, ISSN 1996-1073, www.mdpi.com/journal/energies.

Johnson, G. 2005. A review of bat mortality at wind-energy developments in the United States. *Bat Research News*, Volume 46(2), Pages 45–49.

Kabir, E., Kumar, P., Kumar, S., Adelodun, A., and Kim, K. 2017. Solar energy: Potential and future prospects. *Renewable and Sustainable Energy Reviews*, Volume 82, Pages 894–900. doi:10.1016/j.rser.2017.09.094.

Kaldellis, J., Garakis, K., and Kapsali, M. 2012. Noise impact assessment on the basis of onsite acoustic noise immission measurements for a representative wind farm. *Renewable Energy* Volume 41 (2012), Pages 306e314.

Kaltschmitt, M., Streicher, W., and Wiese, A. 2007. Wind power generation. In: Kaltschmitt M., Streicher W., Wiese A. (eds.) *Renewable Energy*. Springer, Berlin. doi:10.1007/3-540-70949-5_7.

Letcher, T. 2017. *Wind Energy Engineering: A Handbook for Onshore and Offshore Wind Turbines*. Elsevier Science, Amsterdam. ISBN 9780128094297.

MNRE (Ministry of New and Renewable Energy, Government of India). 2021. https://mnre.gov.in/new-technologies/ocean-energy.

National Instruments. 2020. Wind Turbine Control methods. https://www.ni.com/en-in/innovations/white-papers/08/wind-turbine-control-methods.html.

NISE (National Institute of Solar Energy). 2020. 'Surya Rasmi', a monthly newsletter of NISE, Issue 2, Volume 1, December Edition.

Nikitas, G., Bhattacharya, S., Vimalan, N., Demirci, H.E., Nikitas, N., and Kumar, P. 2019. 10-Wind power: A sustainable way to limit climate change, Editor(s): Trevor M. Letcher, *Managing Global Warming*, Academic Press, Cambridge, MA. Pages 333–364. ISBN 9780128141045. doi:10.1016/B978-0-12-814104-5.00010-7.

Pavel, C.C., Lacal-Arántegui, R., Marmier, A., Schüler, D., Tzimas, E., and Buchert, M., et al. 2017. Substitution strategies for reducing the use of rare earths in wind turbines. *Resource Policy* Volume 52, Pages 349–357.

Prasad, R.D., and Bansal, R.C. 2011. Chapter 9 - Photovoltaic systems. *Handbook of Renewable Energy Technology*. Copyright © 2011 by World Scientific Publishing Co. Pte. Ltd. Pages 205–223. ISBN-13 978-981-4289-06-1.

Rao, K.R. 2019. Wind energy: Technical considerations – contents. In: *Wind Energy for Power Generation*. Springer, Cham. doi:10.1007/978-3-319-75134-4_1.

Rivera G., Felix, A., and Mendoza, E. 2020. A review on environmental and social impacts of thermal gradient and tidal currents energy. *International Journal of Environmental Research and Public Health*, Volume 17, Pages 7791. doi:10.3390/ijerph17217791. www.mdpi.com/journal/ijerph.

Seveda, M.S., Rathore, N.S., and Kumar, V. 2011. Chapter 14 - Biomass as a Source of Energy. In: *Handbook of Renewable Energy Technology*. Copyright © 2011 by World Scientific Publishing Co. Pte. Ltd. Pages 323–343, ISBN-13 978-981-4289-06-1.

U.S. Department of Energy Office. 2009. Ocean Energy Technology Overview. U.S. Department of Energy Office of Energy Efficiency and Renewable Energy Federal Energy Management Program, DOE/GO-102009-2823, July 2009. http://large.stanford.edu/courses/2013/ph240/lim2/docs/44200.pdf.

Willis, C.K., Barclay, R.M., Boyles, J.G., Brigham, R.M., Brack Jr., V., Waldien, D.L., and Reichard, J. 2009. Bats Are Not Birds and Other Problems with Sovacools, Analysis of Animal Fatalities Due to Electricity Generation.

Wu, Y., Zhao, F., Liu, S., et al. 2018. Bioenergy production and environmental impacts. *Geoscience Letters* Volume 5, 14. doi:10.1186/s40562-018-0114-y.

Yao, F., Bansal, R.C., Dong, Z.Y., Saket, R. K., and Shakya, J.S. 2011. Chapter - Wind energy resources: Theory, design and applications. In *Handbook of Renewable Energy Technology*. Copyright © 2011 by World Scientific Publishing Co. Pte. Ltd. Pages 3–19. ISBN-13 978-981-4289-06-1.

2 Technical Challenges in Renewable Generations and Their Integration to Grid

D. Blandina Miracle, Rajkumar Viral, and Pyare Mohan Tiwari
Amity University

Mohit Bansal
G L Bajaj Institute of Technology & Management

CONTENTS

2.1 Technical Challenges with the Installations of Renewable Generations 30
    2.1.1 Challenges with Renewable Generations 30
        2.1.1.1 Challenges of Renewable Energy Integration into Grid 33
    2.1.2 Types of Controllers 33
        2.1.2.1 Fuzzy Logic Controller (FLC) 33
        2.1.2.2 SSSC, TCSC, and STATCOM Controllers 33
        2.1.2.3 PI Controller 34
    2.1.3 Advantages 35
    2.1.4 Disadvantages 35
2.2 Some Case Studies of Mitigation of Technical Challenges Achieved 35
    2.2.1 Case Study: Ameren Distribution Microgrid 36
        2.2.1.1 Overview 36
        2.2.1.2 Technical Characteristics 36
        2.2.1.3 Business Model 36
    2.2.2 Shanghai Microgrid Demonstration 37
        2.2.2.1 Overview of Shanghai Microgrid Demonstration Project 37
        2.2.2.2 Technical Characteristics 37
        2.2.2.3 Business Model 38
2.3 Problems Associated with Integration at Transmission and
Distribution Voltage Level 38
    2.3.1 Issues and Challenges 39
    2.3.2 Technical Issues 39

DOI: 10.1201/9781003271857-2

		2.3.2.1	Gird Integration Problems for Small-Scale Generation....... 39
		2.3.2.2	Issues Related to Grid Integration of Large-Scale Generation..39
	2.3.3	Nontechnical Issues 40
2.4	Protection Issues Related to Integration 40
	2.4.1	Protection Issues.. 41
	2.4.2	Possible Solutions to Address Renewable Integration Challenges 42
2.5	Case Studies..42
	2.5.1	Wind Microgrid - Case Study...42
	2.5.2	Case Studies – Barwani 43
	2.5.3	Case Study: San Diego Zoo Solar-to-EV Project 44
		2.5.3.1	Overview of San Diego Zoo Solar-to-EV Project 44
		2.5.3.2	Technical Characteristics..45
		2.5.3.3	Business Model 45
2.6	Conclusion 46
Nomenclature..46
References...47

2.1 TECHNICAL CHALLENGES WITH THE INSTALLATIONS OF RENEWABLE GENERATIONS

Wind and solar photovoltaic systems are growing at a high rate in diverse renewable sources of energy since 2010. The overall world wind power production capability was 487 GW, and the solar PV capacity was 303 GW, analyzing to a penetration point of 4.0% and 1.5% at the end of 2016. Global penetration of renewable energy until December 2016 was just about 8.0%, except traditional hydro share (16.6%). Nevertheless, several countries have set a RE-based electricity generation goal of 30% by 2030. Renewable power generation often normally occurs via distributed generation (DG) (Seck et al. 2020). As a consequence, the architecture of power generation changes from the large, centralized reactors to just a hybrid production pool composed of conventional big plants as well as several minor DG units (Allard et al. 2020). Certain RE generators have different electrical properties to synchronous devices. Even though a wide group of DG systems utilize power electronics converters for equipment, they pose various technological problems relevant to consumer devices, transmission networks, impact generators, control and security and power system activity. Figure 2.1 represents the architectural diagram of the PV system.

2.1.1 CHALLENGES WITH RENEWABLE GENERATIONS

The challenges based on integration can be divided into technological, commercial and regulatory problems that need to be resolved for maximum usage of the RES. The commercial as well as regulatory issues are not listed since only the technical challenges seem to be the focus of attention. The technical problems of integrating large consumption of low PV systems into the delivery network of another grid occur in terms of voltage control, power efficiency, harmonics, and protection difficulties. The reason behind most of the problems was that the distribution network

Technical Challenges in Renewable Integration

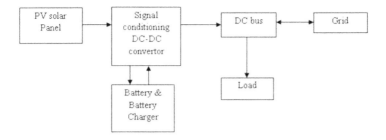

FIGURE 2.1 PV system – architectural diagram.

developed cannot recognize the likelihood of high distribution rates of PV technologies attached and hence modifying the network's process toward unidirectional (passive) to bidirectional (active).

Voltage Regulation: The voltage shift in distribution network is susceptible to short-term variations that may trigger the voltage control devices to break down because of sudden alternations among sunlight and cloud (Chen et al. 2020). In addition, the PV system produced irradiance variability based on the climate changes, and hence, it modifies the voltage performance in PV scheme at PCC, either for a short or long time. The distribution networks in voltage problems owing to the introduction of huge quantity of PV method were categorized as voltage unbalance, voltage flickers, and voltage rise throughout the network.

Power Quality: "Power quality is described as the specification of sensitive equipment for powering, and it is grounded in a way that matches the activity of such equipment". Throughout the description, a common usage of term power quality was established. Power quality generally addresses two significant aspects: the fluctuations of transient voltage as well as the harmonic distortion of network voltage (Zhang et al. 2018). Moreover, in terms of grid and cost, certain problems may occur such as voltage flickers, voltage sag, harmonics, and stress on the supply transformer. The alteration of the network voltage profile and the power flow exchange induced through distributed production seem to be key problems in power quality.

Harmonics Distortion: The harmonic distortion of both the voltage and current waveforms has become a significant problem which needs to be resolved even as PV systems' penetration levels increase throughout the distribution network. Harmonic distortion occurs due to the transformation from DC to AC that is done by an inverter through intrinsic nonlinearity process. Furthermore, the PV system inverters seem to be the key source of current harmonics introduced further into distribution network that might induce harmonic voltage and THD inside the network. Such harmonics lead to a significant increase of losses through heat production in the distribution network.

Protection Challenges: Throughout the distribution network, each feeder's straight protection system requires the feeder to be radial; indicating that the power flow remains unidirectional (Matschoss et al. 2019). While connecting the higher diffusion level of PV method into the distribution system, certain issues should be discussed, including the protection of generation equipment against internal failure, protecting against insulation, and protecting the distribution network against defects (Mendoza-Vizcaino et al. 2019).

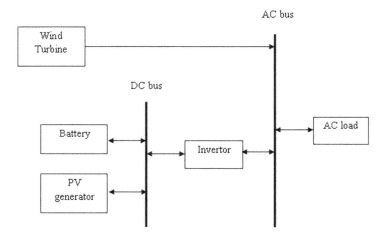

FIGURE 2.2 Block diagram of PV/wind system.

A power plant is really a distributed network which provides electricity to the power grid individually that cooperates with the regulations of the grid system. For improving the efficiency, protection, and grid reliability, the requirements of grid codes are developed to describe the specifications of connecting power plants to the grid and also previously remained focused on traditional power plants (Merai et al. 2020). Originally, the utilization of renewable energies to generate large-scale electricity was really low compared to traditional power plants, although this has significantly changed by allowing the production of grid codes for renewable power plants that needs to prevent impacting grid service. Here, Figures 2.2 and 2.3 demonstrate the architecture of PV/wind system and the grid-connected PV system.

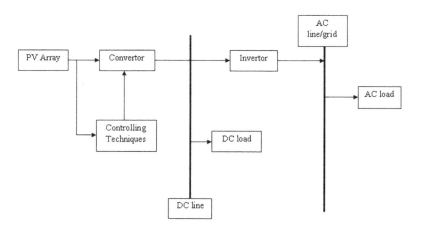

FIGURE 2.3 Block diagram of grid-connected PC system.

Technical Challenges in Renewable Integration

2.1.1.1 Challenges of Renewable Energy Integration into Grid
- The challenge of renewable energy integration into grid is based on the features of renewable resources like uncertainty and variability.
- The above features may affect the techniques used in power system planning and operation.
- It may also add significant costs.

2.1.2 TYPES OF CONTROLLERS

2.1.2.1 Fuzzy Logic Controller (FLC)
The input signals of fuzzy controller's unit were interpreted by the fuzzification unit, and they are assigned with the fuzzy value. The descriptive representation of the variables is regulated by the collection of rules that are focused on process information. The inferential mechanism recognizes the information, taking into consideration certain rules including their membership roles. The defuzzification block is often used to transform to nonfuzzy data that can be used to monitor the procedure of the fuzzy data derived from the inference phase. The designed fuzzy controller is shown below. The presented fuzzy system has two inputs as well as a single output. Figure 2.4 describes the framework of fuzzy logic controller.

2.1.2.2 SSSC, TCSC, and STATCOM Controllers
SSSC of a FACTS system for which the primary purpose would be to adjust the transmission line's characteristic impedance, and to alter the power flow. Transmission line impedance was modified through inserting a voltage that leads or lags the current in the transmission line around 90°. When the SSSC is configured with such an energy storage device, the SSSC was granted the benefit of the power system's reactive and real power compensation. However, the real power was supplied mostly by SSSC with the power storage function through monitoring the angular location

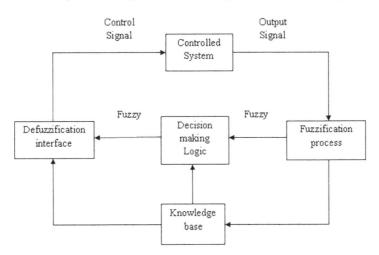

FIGURE 2.4 Block diagram of fuzzy logic controller.

of the injected voltage throughout support of a line current. The lead-lag framework comprising the gain block for the stabilizing agent specifies the quantity of damping. Initially, the washout subblock would be used to decrease the over-response of damping throughout extreme events and acts as a high-pass filter with such a time constant, providing unchanged passage of the signal correlated with rotor speed oscillations. Without this block, modifications to the steady state would alter terminal voltages. Finally, the time constant of its phase compensator component is selected so that it would fully be compensated for the lag/lead phase of a system.

STATCOM is a steady-state operating system that replicates the operational properties of a synchronous compensator which rotates. A STATCOM's basic electronic block is indeed a voltage-sourced converter which transforms a DC voltage at its input terminals into such a three-phase collection of AC voltages with a manageable magnitude and phase angle at frequency components. A STATCOM could be used to regulate voltage in such a power network with the ultimate aim of increasing transmittable power and also improving the features of stable transmission as well as the overall reliability of the device.

2.1.2.3 PI Controller

On the internal control loop, the PI controller controls the DC voltage by contrasting DC link voltage with reference DC voltage. The control loop maintains the DC link voltage since it moves power of the PV network to the grid in a steady and effective way. In RSCAD, each PI controller has two variables to tune, i.e., a proportional gain as well as a time constant (Mohamed et al. 2019). PI controller is also known as feedback control loop which determines the error signal by calculating the variation among the system output, which in this case is the power being drawn from the set point and the battery.

The controller obtains a calculation of both voltage and current that determines the power extracted from battery. Here, the error signal enters PI control loop in which the integral, as well as the proportional constant, multiplies it. The controller output as power value is passed as a power to PWM signal conversion to transform the amount that is equal to control signal. The controlled system component contains the speed controller, motor, battery, and limiter. Figure 2.5 represents the framework of PI controller.

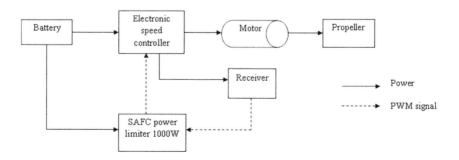

FIGURE 2.5 Block diagram of PI controller.

2.1.3 ADVANTAGES

The major benefits of a grid-connected PV network would be its efficiency, fairly small operational and maintenance costs, including decreased bills for electricity. The off-grid network has the benefits of even being fully energy-independent, ensuring that it might not have the limitations that come through utilizing the main electric grid, like power outages (Eissa 2018). An off-grid network will practically be installed somewhere where there is light.

2.1.4 DISADVANTAGES

- The drawback is that it would be essential to develop a required number of solar panels for producing the needed quantity of excess power.
- Installations on the ground are sporadic and unpredictable. The amount of electricity produced depends mostly on climatic conditions. These could not produce electricity at night or in rainy days or occasionally.
- Low energy density. The duration of the solar radiation obtained only at the ground during normal conditions becomes $1,000W/m^2$. This tries to cover a wide area if used in large size.
- The price is also relatively high; the cost of installation is high, 3–15 times than the traditional energy production.

2.2 SOME CASE STUDIES OF MITIGATION OF TECHNICAL CHALLENGES ACHIEVED

Wind power generation seems to be an innovative source of renewable energy produced via WECS. Although WECS has various features from traditional power plants, its incorporation into another power system grid leads to considerable technical problems with regard to power quality (Ju et al. 2020).

For a small time frame, the instantaneous peak power that would have been produced from its PV array is assumed to be constant, in which the distribution network is exposed to maximum potential consequences (extreme impacts). Subsequently, the PV array generates the peak (extreme) power according to the rating of the mounted PV array in which the higher impacts were obtained. The effects from both PV size (penetration level) and PV position (sitting) cause many technical problems including voltage profile across various buses, power losses throughout cables, and short-circuit currents across various nodes that are susceptible to symmetrical fault currents (Gur et al. 2018; Shahnazian et al. 2018).

The solar and wind production occurs due to instability and uncertainty that might present difficulties to grid operators. In order to manage supply-side volatility and the link between levels and loads of generation, greater flexibility in the network may still be required (Jafari et al. 2019). Some other problem occurs when the wind or solar energy is at low load levels; however, traditional generators might have to shut down in low generation levels of certain cases (Cifor et al. 2015).

2.2.1 CASE STUDY: AMEREN DISTRIBUTION MICROGRID

Project Background: The microgrid of Ameren located at Engineering Applications Center in Champaign, Illinois, operates at an energetic useful transmission feeder as well as, through combination with the main grid, and could deliver above 1 MW of electricity to limited consumers, increasing efficiency and stability. This is one among the several microgrids that work on utility/medium voltages, is the first utility microgrid operated by developers in island consumers, but utilizes the special control network for cyber-secure microgrid as well as intelligent automation scheme for smooth transformation of microgrid mode returning to grid-connected mode.

2.2.1.1 Overview

Location: Champaign, Illinois, Capacity: 1.475 MW, Utility: Ameren Corporation (Ameren Illinois), Total Cost: Approximately $5 million, Host Organization: Ameren Technology Applications Center, Installation Timeline: June–December 2016, Developer/Vendor(s): Ameren, Schneider Electric, S&C Electric Company, Systems, Yingli, Northern Power Caterpillar.

S&C Electric Company equipped Ameren with a full EPC solution, including layout, technology, and interconnection analyses, thus providing the cyber-secure device, power storage, communications network, and delivery switchgear as well as other appliances.

2.2.1.2 Technical Characteristics

- 125 kW solar PV, Yingli
- Microgrid : Schneider Electric, Economic Advisor
- ABB, String inverters.
- S&C Electric subsidiary IPERC, GridMaster® military-grade cyber-secure controller.
- Northern Power Systems, 100 kW wind turbine.
- S&C Electric Company
- Caterpillar, 1 MW (2×500 kW) natural gas generators.
- Costs: Approximately $5 million.

2.2.1.3 Business Model

Financing: Funding for the research was authorized by the "Energy Infrastructure Modernization Act" that allowed Ameren Illinois to contribute $643 million over the 10-year program's existence to revitalize the energy distribution network.

Value Proposition: The project should include the Ameren's primary factors such as stronger customer reliability, improved knowledge of DER application, enhancing consumer energy quality, and ensuring a constant supply of energy during insulation events. Ameren wishes to offer its consumers certain desired advantages.

Revenue Streams: As this microgrid is used for the delivery of services, microgrid and hence no revenue streams were produced throughout operations for any third party.

Lessons Learned: To complete this project S&C Electric Company is an extremely short timeline. Since Ameren Illinois would be a distribution-only utility, to keep the project industry acceptable, S&C Electric Company decided to find a hire business

Technical Challenges in Renewable Integration 37

based on generation machines. Major key point would be that the regulatory climate would have a huge effect mostly on the project plan, the project timeline, and the financing scheme.

The project offered a range of technological lessons as well. S&C Electric Company also introduced modern control systems and new frequency modulation and voltage management protection systems that uses two cases of interest in Ameren. Such skills were especially essential for optimizing a device with intermittent microgrid generation to prevent wasteful activity of the natural gas producers.

2.2.2 SHANGHAI MICROGRID DEMONSTRATION

Project Background: This research involves a partnership between the University of Tsinghua in China and the University of Aalborg in Denmark. Here, this project established a "research-oriented, demonstrative platform, which plans to relieve combining DG units, multilevel control strategies and hierarchical, as well as numerous configurations of microgrid with the aid of communication scheme on the basis of standard, which results in whole solutions of microgrid" (Howell, et.al., 2017).

The demonstration will build, evaluate, and assess the efficiency of the microgrid as well as the output of individual components in the integrated device to achieve its goals. Important elements for this project include AMI as well as smart grid systems.

2.2.2.1 Overview of Shanghai Microgrid Demonstration Project

Location: Shanghai, China, Capacity: 206 kW, Utility: STATE GRID Corporation of China, Host Organization: Plant B HT-Shanghai Solar, Total Cost 2,300,000 DKK26 ($371,400 USD at a 0.16 conversion), Installation Timeline: September–December 2013 installation; 2014–2017 commissioning as well as testing, Developer/Vendor(s): Aalborg University, Tsinghua University, Kamstrup A/S, Shanghai SolarEnergy Science & Technology Co.

Tsinghua University has built the 200 kW microgrid processing plant in Shanghai Solar in collaboration with Kamstrup A/S and Shanghai Solar. Scientists are modeling the microgrid established in Shanghai at the laboratory at Aalborg University. They are operating to improve the management system for the AMI and also for resources.

2.2.2.2 Technical Characteristics
- EMS
- 136 kW rooftop solar PV
- 50 kW energy storage system
- 20 kW (2×10 kW) wind turbines
- Costs: Public funding offered 1,000,000 DKK as well as private funding offered 1,300,000 DKK for the project (56.5% private). The expenditure of the project was $371,400 USD or 2,300,000 DKK at a conversion rate of 0.16 in September 2017.
- AMI and smart grid communication technology

2.2.2.3 Business Model

Financing: The revelation project was collectively funded via HT-Shanghai Solar (1,300,000 DKK) as well as the RED in China and Denmark (1,000,000 DKK).

Value Proposition: The approach presented would promote both planning and organizational research. Robust reliability and power efficiency should also be recognized by careful monitoring of the interfaces of power electronics as well as the integration of AMI including "smart grid ready networking technologies".

Revenue Streams: The researchers built a cost-based revenue model as one of the implementation project to analyze and calculate the microgrid ROI. The ROI is measured focused on the microgrid's performance and operating existence. The microgrid produced revenue from its electricity sales to a utility in operation.

Lessons Learned: Operationally, it was able to coordinate over great distances, and time change seemed to be essential for the project collaborators. A more significant lesson is the value of preparation over time for adjustments. Changes in procedures and infrastructure are discussed in detail throughout the project preparation process.

2.3 PROBLEMS ASSOCIATED WITH INTEGRATION AT TRANSMISSION AND DISTRIBUTION VOLTAGE LEVEL

Integrating wind and solar energy through conventional power systems poses technological problems such as voltage control, flickering, harmonic distortion, stability, etc. (Komiyama et al. 2015). Such issues of power quality need to be limited with IEC and IEEE standards. An analysis of several recent years' papers reveals that such issues of power quality will arise at generation, transmission, and distribution.

Voltage Regulation: The droop features have been used to regulate the voltage magnitude and frequency, especially for DFIGs. This could be applied to WECS by making an analysis of voltage reliability to accomplish voltage control of PCC. Because of its voltage-drive function, the high DC bus ripple provides better AC power quality and suggests a bidirectional power flow as well as integrated bottom-up control methods enable DG method with better regulated organized values (Akhtar and Saqib 2016). Solving that issue with fuzzy logic controllers in the grid-interfacing system will increase the voltage efficiency (Luhmann et al. 2015).

Voltage Sags/Swells: The grid-connected sensitive load activity is affected by the voltage drops. The power electronic converter is introduced through the usage of a series compensator to address this drawback that the author requires a reduced amount of active power significantly and seems to recover the load-side voltage (Hassan et al. 2015). "Grid-interfacing power quality compensator including 3-phase 4-wire microgrid applications were built with sequence components for insert voltages as just a corresponding metrics in the Net-metering scenarios a PQCC could regulate voltage because of the reversal of DG power flows and thus the raise in short-circuit current" (Li, et. al., 2006).

Harmonics: In past few years, wind turbines have concentrated on grid activity and grid effects. The reason for this concern would be wind turbines as possible sources of poor power supply among the distributors. In particular, a wind turbine

Technical Challenges in Renewable Integration

with different speed provides other benefits in flicker. But, with different speed wind turbines, a new issue occurred. "Modern forced-commutated inverters" were used throughout the wind turbines with variable speeds not just to produce inter harmonics as well as thereby harmonics.

Real and Reactive Power: Regarding GCWT networks, the weather conditions and the diurnal wind fluctuations should have to be addressed to produce high-quality power through inverters which follow the grid codes requirements. "A droop control system is suggested depending on the reactive power generated through the current of a negative sequence and the voltage of a positive sequence line" (Zhang et al. 2017). The current monitoring controller based on the Lyapunov feature was implemented for controlling the reactive and real power flow also for the parallel-connected inverter.

Power System Stability Problem: The stability of the power system is divided into stability of the rotor (or strength) angle; (ii) stability of the frequency; and (iii) stability of the voltage. Instability would lead to higher angular swings of certain generators resulting in the synchronization loss in many generators (Rahbar et al. 2015). The voltage stability is associated with the capacity of a power plant to sustain its steady voltage throughout all buses in the network at normal operating conditions even after disturbances (Luo et al. 2018).

2.3.1 Issues and Challenges

A renewable energy source was nonlinear in the environment, and incorporating sustainable energy resources through another power grid is therefore a difficult challenge (Parizy et al. 2020). Many of the challenges and problems are involved with grid integration of different renewable energy sources, especially solar photovoltaic and conversion systems in wind power. These challenges are also widely divided as nontechnical and technical as given as follows.

2.3.2 Technical Issues

2.3.2.1 Gird Integration Problems for Small-Scale Generation

Cost, grid congestion, bidirectional power flow in the distribution network, variability of renewable production, protection issues, weak grids, low power quality localized, and voltage stability issues have challenged reliability and efficiency of grid interface.

2.3.2.2 Issues Related to Grid Integration of Large-Scale Generation

Transient stability, voltage stability, small disturbance power system stability, frequency, large disturbance, rotor angle, angle stability, stability, and voltage stability are certain aspects to be considered while modeling a grid system. The recent improvement in wind energy generation has developed major wind farms with capability exceeding 100 MW. In fact, these large wind turbines were interconnected to the grid.

The reactive power requirement is the main problem in wind power generation for voltage support that includes power quality problems consisting of synchronizing and starting of wind farms to the grid, turbine power electronic design and controller optimization, voltage flicker, sub-synchronous resonance problems based on the complex shaft/gear system of wind turbine, the problems of wind farms connected into series compensated systems, and interaction of the electric network.

2.3.3 NONTECHNICAL ISSUES

Lack of nonconstruction of new reserve power plants, less availability of transmission line, and technical skilled manpower to accommodate RES are some of the nontechnical issues in grid-connected systems. Long-term, idealistic renewable energy goals create a framework which increases fuel in sustainable energy policy and systems integration. "Grid-aware" rewards are indeed essential for promoting innovation in renewable energy and reducing the negative impacts of introducing such services into the power system.

2.4 PROTECTION ISSUES RELATED TO INTEGRATION

Growing with the RES and transmitted generators involve additional strategies for operating and managing the electricity grid to preserve or even enhance the stability and performance of the power supply (Yi et al. 2018). The researchers have implemented some solutions as determined below.

1. Power electronic infrastructure is a significant role in DG, the implementation of RES into the electricity grid and also that technologies are increasingly interconnected with grid-based networks; it's also commonly utilized and expanding quickly. Power electronics has experienced a rapid transformation over the past few years, owing mainly to two reasons (Chen et al. 2016). Grid-friendly converters and cost efficient production have contributed to this transformation.
2. The RES intermittence of power output may be regulated as extracting the power through spreading in small units of wider geographical region rather than a huge centralized unit in one location. The fluctuation of overall output power could be reduced since low unit power can be influenced by the local problems.
3. The load is fed mostly in off-peak load time during nighttime hours, while irrigation is powered, and it is supplied by a traditional grid. At the other side, RES powers such as solar PV are produced during the daytime, and electricity for irrigation purposes rather than earlier accumulates the energy that reduces the overall system costs.
4. The plant power output trends in large solar PV vary throughout the day when the whole power was connected to the grid that constantly fluctuates power leads to the grid's security issue (Nan et al. 2019). The developer of the solar PV plant should install various kinds of storage devices that give the additional costs for plant owner.

2.4.1 PROTECTION ISSUES

The key security problems relevant to distributed grid power connection are a) change of short-circuit levels, b) reverse power flow, c) lack of sustained fault current, d) blinding of protection and e) islanding.

Change in Short-Circuit Levels: Direct synchronous machine-based DRs attached to the grid raise the level of network failure. Induction generators could have a limited amount of fault currents, not in a sustainable way. A high fault level, as stated earlier suggests a strong grid in which the possible consequences of introducing a new DG to the PCC voltage would not be quite significant. Nevertheless, the rise in the level of network faults due to the introduction of DG reflects another system security problem. The short-circuit level at fault point is described by the equivalent system displacement. The corresponding network impedance will reduce the attachment of generators, leading to an increase in the fault frequency. In this scenario, the fault current can exceed the ability to exist CBs to split. In addition, the modified fault levels can interfere with communication between overcurrent relays culminating in the inadequate activity of protection systems. This would also significantly raise losses and lead to larger differences in voltage only at the generator.

Reverse Power Flow: Distribution networks are generally radial structures and are typically covered against established security schemes utilizing time grading in reverse power flow process. Some protections of the distribution system recognize an abnormal condition by distinguishing a current of fault from the usual state of the charge.

Lack of Sustained Fault Current: "For the safety relays to reliably distinguish and differentiate fault currents again from usual load currents, the faults should lead to an important and persistent rise in the currents determined through the relays" (https://www.iitr.ac.in/departments/HRE/uploads/modern_hydroelectric_engg/VOL_2/Chapter-3_Electrical_Protection_System.pdf). Whenever a DG's current contributions to fault are small, it is challenging for the overcurrent safety relays to detect faults efficiently. Production derived from renewable energies also utilizes induction generators, electronic converters, or smaller synchronous generators of energy. Induction generators could not provide three-phase faults with continuous fault currents, and therefore can make a simple connection to asymmetric faults. The exclusion of current sustained fault cooperate the capacity of circuits to identify faults.

Blinding of Protection: The grid relation to the overall fault conditions would decrease in the presence of DG. Because of this significant decrease, short-circuit may remain undetected, as the grid contribution to the short-circuit current doesn't ever attain the feeder relay's pickup current. Protections based on such devices can therefore suffer from malfunction due to decreased grid contribution.

Islanding: Islanding seems to be the process of disconnecting a portion of the network from the main grid and functioning as an isolated facility equipped with one or many generators. Islanding leads to the irregular frequency and voltage differences throughout the insulated network. In fact, islanding activity can generate an ungrounded network depending on the connections to the transformer. Islanding is regarded an unsafe situation, and therefore, it is recommended that the DGs be immediately disconnected from the main grid of an island.

2.4.2 POSSIBLE SOLUTIONS TO ADDRESS RENEWABLE INTEGRATION CHALLENGES

The biggest concern in considering a technical approach is the cost-effectiveness of the system and the network characteristics. Grid technology, operating activities, type of generation, and regulatory aspects all affect the most feasible and viable methods of solutions. Reliability can be accomplished through better forecasting, operating practices, energy conservation, versatility on demand, flexible generators, and other mechanisms.

Forecasting of Wind and Solar Resources: Solar and wind forecasting could even help decrease the uncertainty associated with those generations. There are various forms of forecasting, including such short- and long-term forecasting. The short-term forecast, typically in hours, becomes considerably less complicated than the long-term forecast. Usually, prediction errors vary from 3% to 6% of the estimated potential 1 hour ahead as well as 6%–8% a day ahead.

Operational Practices: Wider Balancing Authority Areas and Quick Dispatch: Rapid dispatch aids to control RE power volatility since it eliminates the need for resource management, enhances effectiveness, and offers exposure to a wider set of resources to support the system. For quicker dispatch, loading and production rates could be more closely adjusted, reducing the need for further costly resources for regulation.

Reserves Management: Modified resource management practices could be used to help tackle wind and solar volatility. It involves (1) restricting wind and photovoltaic ramps to minimize reserve requirements and (2) requiring intermittent renewable sources to provide reserves or other ancillary services, including such oversight, inertia, etc.

Interconnecting More Distributed Resources: The impacts of RE power intermittent could be reduced by interconnecting vast amounts of small scattered resources distributed across a wider geographic region rather than focusing big units in one location. The overall output fluctuations should be minimal because the local differences only impact specific units, not the entire output power.

Energy Storage: For rising renewable penetration levels, energy storage is a typical approach to mitigate the restriction of production. Major "overbuilding" (200%–300%) and curtailment provide an alternative to costly storage structures.

Wind-PV Hybrid Systems: Considering the wind and solar PV outputs are complementary forms, the hybrid configuration of such two resources would to some extent increase the overall power fluctuations.

Demand Response: Demand-side flexibility is a strong choice for raising the impacts of quick ramps. The response to demand could be used for reserves and ancillary services and also peak reduction. Using the response of demand to maintain the stability of the system in rare occurrences where there is a significant under or over-supply of renewable generation may cause cost savings when considered with continuous maintenance of extra reserves.

2.5 CASE STUDIES

2.5.1 WIND MICROGRID - CASE STUDY

There are more researches carried out from different countries to expertise the case study concerning microgrids. Through the implementation of distributed technology, there seems to be a number of case studies beyond India. The most successful microgrid case study is the wind microgrid of Etechwin Microgrid Solutions.

Technical Challenges in Renewable Integration 43

Here, two 1.5 MW production wind turbines and both super capacitors in a specialized pitch system in the microgrid network are considered. The usage of super capacitors in the photovoltaic device generates electricity, and it is often fed into energy storage networks at the park level. The EMS comprises of control methods for regulating the active IGBT system and the wind turbine pitch control as well as the speed regulation. The voltages should be calculated by the EMS in relation to the strength of the AC bus, and the control behavior should be weighed. The production, storage, and control supplies are achieved entirely by wind community and Etechwin Solutions. This microgrid device is situated in a land area of 91,271 m^2 at Beijing's wind headquarters and has a production value of 200—2,200 kW. The rating of every source utilized in the microgrid is described as follows.

Wind Turbine: GW109/2,500 kW type and height of 80 m. Micro-gas turbine: Cogeneration micro turbine production with two turbines of 65 kW. Solar photovoltaic: 500 kW capacity = polycrystalline silicon 490 kW, single crystal silicon 5 kW, and cadmium telluride thin film 5 kW. Energy storage system: Vanadium flow of 200 kW*4 hr, lithium battery 200 kW*4 hr, supercapacitor of 200 kW*10 sec.

The implementation of supercapacitors incorporates and maximizes the proportion of RE production in the smart microgrid energy management network. It will organize the entire network in insulated and grid-connected service which ensures PS stability. In the research scenario, they use an adaptive control interface to monitor the different technologies of production, loading, and storage. Along with system operation, the control approach allows for the personalized and flexible specification. The following facilities given by the wind microgrid are fully automated microgrid, high-speed information monitoring including power flow, power resources status in each mode of operation, lower cost, improvement in power quality, higher speed event recording, and the ability for control of bidirectional PS.

2.5.2 Case Studies – Barwani

This case study, introduces optimal sizing of diverse variations of energy-based hybrid wind and PV power systems throughout the central area by means of HOMER software tool.

Renewable Energy Resources: A Jamny Ven village Barwani (longitude 75.85 as well as latitude 22.71) Madhya Pradesh, a platform of renewable energy located in India, seems to be an important component of hybrid systems. As per IMD, significant amounts of wind and solar resources were essential in several areas of India. Such sources of electricity are irregular and achievable naturally. Environment data for different site hybrid RESs are an essential factor in the analysis of the former's capacity for sensitive details, and wind and solar energy statistics are taken from NASA.

Solar Energy Resource: Hourly information regarding emission of solar levels was gathered from the village of Barwani Jamny ecosystem. Annual estimation of long-term resource scaling is (5,531). In the summer season, solar power is higher compared to the winter season.

Wind Energy Resource: The wind energy supplies of the mechanisms of modification seem to be pollution-free and are readily available. Periodic frequent wind data for the village of Jamny Ven were finally starting along with the Barwani climate system. The scaled average of yearly wind speed was 4.5 m/s; the peak monthly average wind speed is measured with such a higher value of 7.195 m/s mostly during the month of December as well as the lowest value is recorded with a monthly average wind speed of 2.664 m/s throughout June.

Electrical Load Data: The daily consumption rate of average energy is 110.6 (kWh/day), the peak load is 13.23kW, as well as the average is 4.61kW. The data have been measured as the normal load of electrical situation of a market for just a village in district of Barwani.

The project aims to develop optimum sizing of various mixtures of "energy-based hybrid power system for electrification of rural main region (Jamny Ven Barwani)" in which the expense of utility supply is very soaring based on small demand superior transmission as well as high transport costs. The case study selected has a power demand of 110.6 kWh/d. The two hybrid systems are approximately identical in pollution and levelized COE and hence the "overall NPC and running costs of the PV – Wind – Battery – DG" are smaller than those of the wind-DG hybrid device. If the solar penetration decreases, the wind power rises; the carbon surplus multiplies. Besides allowing use of the battery bank, it can be protected and used by predictable future purpose.

2.5.3 CASE STUDY: SAN DIEGO ZOO SOLAR-TO-EV PROJECT

Project Background: The Solar-to-EV project, a collaborative project among San Diego Zoo as well as Smart City San Diego, is formally renowned as the San Diego Zoo Microgrid. Smart City San Diego is indeed a private and public partnership which involves the City of San Diego, SDG&E, GE, and Clean TECH San Diego including San Diego California University. The aim of the initiative is to "enhance the region's energy choice, to allow customers to utilize electric vehicles, to diminish emissions of greenhouse gas, as well as to promote growth of economy." The project of Solar-to-EV involves five charging stations of solar-to-EV driven by a 90-kW solar array throughout 10 canopies of parking lots".

2.5.3.1 Overview of San Diego Zoo Solar-to-EV Project

Location: San Diego in California, Host Organization: San Diego Zoo /Smart City San Diego, Capacity:190kW, Utility: SDG&E, Total Cost: $1,000,000, Installation: Timeline 1year (2012) together with a development period of 6months and Developer/Vendor(s): SDG&E, Independent Energy Solutions, Kokam, Princeton Power Systems, Kyocera, ECOtality (https://www.districtenergy.org/HigherLogic/System/DownloadDocumentFile.ashx?DocumentFileKey=ece11ac1-15f2-0be9-4cf3-8ed53d6260d3).

The solar-to-EV initiative offers essential charging services, highlights the dedication of San Diego to renewable energies, and helps to improve the environmental quality in San Diego. Independent Energy Solutions acted as the project contractor and took prominent architecture and construction architects, engineers, and designers to San Diego. SDG&E manages and controls the system, with advantages to San Diego rate payers of the electric grid.

2.5.3.2 Technical Characteristics

- 100 kW / kWh Li-ion polymer batteries (2×50 kWh battery banks), Kokam
- Costs: The project total expenditure was $1,000,000 in 2012.
- 90 kW solar PV parking array (420×245-watt panels) on 10 stand-alone canopies, Kyocera
- Princeton power systems, bidirectional inverters.
- Five Level 2 EV charging stations, ECOtality Blink

2.5.3.3 Business Model

Financing: The project was sponsored by SDG&E as a platform for the company to evaluate and verify EV chargers, distributed power, dispatchable energy storage, as well as insulating portions of the electrical network in order to enhance reliability under the condition of transient network. The effective orientation project is operated and controlled by SDG&E.

Value Proposition: The community value proposition is consistent with Smart City San Diego's priorities of delivering public EV charging, reducing energy consumption and greenhouse gas emissions, helping small enterprises, and promoting the city's leadership in clean energy. It is also estimated that "project would decrease pollution of carbon dioxide by 189,216 pounds, equal to the planting of 2,788 trees that develop for ten years or eliminating 21 vehicles from road per year" (https://theicct.org/sites/default/files/publications/EV-life-cycle-GHG_ICCT-Briefing_09022018_vF.pdf). The primary benefit of the solar-to-EV initiative for the company and its rate payers is a first-of-a-kind test ground for active regulation of utility for smart microgrids. "This integrates energy production, charging grid efficiency, reduction of DG, and peak demand by hierarchical controls" (https://www.niti.gov.in/sites/default/files/2021-09/Report1-Fundamentals-ofElectricVehicleChargingTechnology-and-its-Grid-Integration_GIZ-IITB.pdf).

Streams of Revenue: "The San Diego Zoo collects the ongoing certificate charge from SDG&E to operate the device" (https://docs.cpuc.ca.gov/published/Final_resolution/22277.htm).

Lessons Learned: "The Solar-to-EV project introduces better public and private partnership, creative innovation in DER implementation by attaching solar PV to EV charging, and links with the smart city idea. The companies owned by some other investor are Electric and Baltimore Gas, PEPCO, as well as Commonwealth Edison, also sought to pay microgrids; however, state regulators and/or legislator regulations often resisted these plans" (https://www.districtenergy.org/HigherLogic/System/DownloadDocumentFile.ashx?DocumentFileKey=ece11ac1-15f2-0be9-4cf3-8ed53d6260d3).

This project presents an illustration in which the link to San Diego's targets with the idea of "smart city" is specifically associated in microgrid. The link in sustainable energy was considering that 40% of California's carbon emissions come from the transportation industry. The concept of a smart city is larger than electricity. Microgrids should be able to facilitate the growth of smart cities in the future by incorporating numerous DERs, like EVs, into microgrid networking systems.

2.6 CONCLUSION

In this chapter, we have made a discussion on technical challenges in renewable generations and their integration to the grid. Moreover, some case studies of the mitigation of technical challenges have been analyzed. This chapter also defines the problems associated with integration at transmission and distribution voltage levels. Finally, the protection issues related to integration have also been studied in this chapter.

NOMENCLATURE

Abbreviation	Description
ANF	Adaptive Notch Filtering
AC	Alternating Current
AMI	Advanced Metering Infrastructure
CoE	Costs Of Energy
DC	Direct Current
DER	Distributed Energy Resources
DEA	Differential Evolution Algorithm
DG	Distributed Generation
EV	Electric Vehicle
EIM	Energy Imbalance Market
EMS	Energy Management System
EMP	Electromagnetic Pulse
EMS	Energy Management System
FLC	Fuzzy Logic Controller
GA	Genetic Algorithm
GW	Giga Watts
GCWT	Grid-Connected Wind Turbine
IEC	International Electrotechnical Commission
IMD	India Meteorological Department
IGBT	Insulated-Gate Bipolar Transistor
KW	Kilo Watts
LV	Low Voltage
MW	Mega Watts
NASA	National Aeronautics And Space Administration
NPC	Net Present Cost
PCC	Point Of Common Coupling
PWM	Pulse Width Modulation
PV	Photo Voltaic
PI	Proportional Integral
PQ	Power Quality
PSO	Particle Swarm Optimization
PQCC	Power Quality Control Center
RED	Renewable Energy Development Program

(Continued)

Abbreviation	Description
RES	Renewable Energy Sources
RE	Renewable Energy
ROI	Rate-Of-Return
SDG&E	San Diego Gas and Electric
THD	Total Harmonic Distortion
WECS	Wind Energy Conversion Systems
WWSIS	Western Wind And Solar Integration

REFERENCES

Akhtar, Z., Saqib, M. A. 2016. Microgrids formed by renewable energy integration into power grids pose electrical protection challenges. *Renew Energ* 99: 148–157.

Allard, S., Mima, S., Debusschere, V., Quoc, T. T., Hadjsaid, N. 2020. European transmission grid expansion as a flexibility option in a scenario of large scale variable renewable energies integration. *Energ Econ* 87: 104733.

Chen, H., Xuan, P., Wang, Y., Tan, K. and Jin, X. 2016. Key technologies for integration of multitype renewable energy sources: Research on multi-timeframe robust scheduling/dispatch. *IEEE T Smart Grid* 7(1): 471–480. doi:10.1109/TSG.2015.2388756.

Chen, T., Pipattanasomporn, M., Rahman, I., Jing, Z., Rahman, S. 2020. MATPLAN: A probability-based planning tool for cost-effective grid integration of renewable energy. *Renew Energ* 156: 1089–1099.

Cifor, A., Denholm, P., Ela, E., Hodge, B.-M., Reed, A. 2015. The policy and institutional challenges of grid integration of renewable energy in the western United States. *Util Policy* 33: 34–41.

Eissa, M. M. 2018. New protection principle for smart grid with renewable energy sources integration using WiMAX centralized scheduling technology. *Int J Elec Power* 97: 372–384.

Gur, K., Chatzikyriakou, D., Baschet, C., Salomon, M. 2018. The reuse of electrified vehicle batteries as a means of integrating renewable energy into the European electricity grid: A policy and market analysis. *Energ Policy* 113: 535–545.

Hassan, H. A. H., Pelov, A. and Nuaymi, L. 2015. Integrating cellular networks, smart grid, and renewable energy: Analysis, architecture, and challenges. *IEEE Access* 3: 2755–2770. doi:10.1109/ACCESS.2015.2507781.

Howell, S., Rezgui, Y., Hippolyte, J. L., Jayan, B. and Li, H. 2017. Towards the next generation of smart grids: Semantic and holonic multi-agent management of distributed energy resources. *Renewable and Sustainable Energy Reviews* 77: 193–214.

Jafari, M., Malekjamshidi, Z., Zhu, J. 2019. Design and development of a multi-winding high-frequency magnetic link for grid integration of residential renewable energy systems. *Appl Energ* 242: 1209–1225.

Ju, L., Tan, Q., Lin, H., Mei, S., Wang, Y. 2020. A two-stage optimal coordinated scheduling strategy for micro energy grid integrating intermittent renewable energy sources considering multi-energy flexible conversion. *Energ* 196: 117078.

Komiyama, R., Otsuki, T., Fujii, Y. 2015. Energy modeling and analysis for optimal grid integration of large-scale variable renewables using hydrogen storage in Japan. *Energ* 81: 537–555.

Li, Y. W., Vilathgamuwa, D. M. and Loh, P. C. 2006. A grid-interfacing power quality compensator for three-phase three-wire microgrid applications. *IEEE transactions on Power Electronics* 21(4): 1021–1031.

Luhmann, T., Wieben, E., Treydel, R. Stadler, M., Kumm, T. 2015. An approach for cost-efficient grid integration of distributed renewable energy sources. *Eng* 1(4): 447–452.

Luo, C., Huang, Y. and Gupta, V. 2018. Stochastic dynamic pricing for EV charging stations with renewable integration and energy storage. *IEEE T Smart Grid* 9(2): 1494–1505. doi: 10.1109/TSG.2017.2696493.

Matschoss, P., Bayer, B., Thomas, H., Marian, A. 2019. The German incentive regulation and its practical impact on the grid integration of renewable energy systems. *Renew Energy* 134: 727–738.

Mendoza-Vizcaino, J., Raza, M., Sumper, A., Díaz-González, F., Galceran-Arellano, S. 2019. Integral approach to energy planning and electric grid assessment in a renewable energy technology integration for a 50/50 target applied to a small island. *Appl Energy* 233–234: 524–543.

Merai, M., Naouar, M. W., Slama-Belkhodja, I., Monmasson, E. 2020. A systematic design methodology for DC-link voltage control of single phase grid-tied PV systems. *Math Comput Simulat* 183:158–170.

Mohamed, I., Mosaad, M., Abed El-Raouf, O., Al-Ahmar, M. A. Bendary, F. M. 2019. Optimal PI controller of DVR to enhance the performance of hybrid power system feeding a remote area in Egypt. *Sustain Cities Soc* 47: 101469.

Nan, J., Yao, W. and Wen, J. 2019. Energy storage-based control of multi-terminal DC grid to eliminate the fluctuations of renewable energy. *J Eng* 2019(16): 991–995. doi: 10.1049/joe.2018.8440.

Parizy, E. S., Choi, S. and Bahrami, H. R. 2020. Grid-specific co-optimization of incentive for generation planning in power systems with renewable energy sources. *IEEE T Sustain Energy* 11(2):947–957. doi:10.1109/TSTE.2019.2914875.

Rahbar, K., Xu, J. and Zhang, R. 2015. Real-time energy storage management for renewable integration in microgrid: An off-line optimization approach. *IEEE T Smart Grid* 6(1): 124–134. doi:10.1109/TSG.2014.2359004.

Seck, G. S., Krakowski, V., Assoumou, E., Maïzi, N., Mazauric, V. 2020. Embedding power system's reliability within a long-term energy system optimization model: Linking high renewable energy integration and future grid stability for France by 2050. *Appl Energy* 257: 114037.

Shahnazian, F., Adabi, J., Pouresmaeil, E., Catalão, J.P.S. 2018. Interfacing modular multilevel converters for grid integration of renewable energy sources. *Electr Pow Syst Res* 160: 439–449.

Yi, W., Zhang, Y., Zhao, Z., Huang, Y. 2018. Multiobjective robust scheduling for smart distribution grids: Considering renewable energy and demand response uncertainty. *IEEE Access* 6: 45715–45724. doi:10.1109/ACCESS..2865598.

Zhang, D., You, P., Liu, F., Zhang, Y., Feng, C. 2018. Regulating cost for renewable energy integration in power grids. *Global Energ Interconnect* 1(5): 544–551.

Zhang, Y., Dai, X. and Han, X. 2017. Renewable energy integration capacity assessment in regional power grid based on an enhanced sequential production simulation. *J Eng* 2017(13): 1065–1070. doi:10.1049/joe.2017.0493.

3 Renewable Energy Integration Issues from Consumer and Utility Perspective

K. Pritam Satsangi, G. S. Sailesh Babu,
Bhagwan Das Devulapalli, and Ajay Kumar Saxena
Dayalbagh Educational Institute

CONTENTS

3.1 Introduction 50
    3.1.1 Literature Review – Grid Integration Issues 50
3.2 Performance Parameter Issues to Consumer and Utility –
    Performance Issues 52
    3.2.1 Methodology 53
        3.2.1.1 Indian Standards 53
        3.2.1.2 Description of the SPV MicroGrid (SPVMG) 53
        3.2.1.3 Measured Parameters 53
        3.2.1.4 Derived Parameters 55
3.3 Performance Results and Issues at the Consumer End 58
3.4 Mitigation of Issues Due to Grid Integration 65
3.5 Financial Implications to Consumer/Utility 67
3.6 Case Study 68
    3.6.1 Major Revisions in IEC 61724 (Standards) 68
    3.6.2 New Performance Metrics 68
        3.6.2.1 Soiling Ratio 68
        3.6.2.2 Soiling Level 70
        3.6.2.3 Performance Ratio 70
        3.6.2.4 Temperature-Corrected Performance Ratio 70
        3.6.2.5 STC-Corrected Performance Ratio 72
        3.6.2.6 Discussion and Results 72
3.7 Conclusions 75
References 76

DOI: 10.1201/9781003271857-3

3.1 INTRODUCTION

Integration of renewable energy systems with the grid has always been a challenge since the inception of such systems. Initially, renewable systems were standalone type with few advantages for local energy support. Integration with the grid has brought up new challenges with an increased scope of research across the world. Grid integration of renewable energy deals with an efficient way of delivering generated energy to the grid. One of the major outcomes of grid integration studies or research is better planning of future renewable energy installations and their utilization (NREL, 2013). The Greening The Grid (GTG) initiative by US-India Collaboration is a one-of-its-kind grid integration study and analysis of renewable systems. Optimization of power system with increased solar/wind integration and strategies for improved grid integration are two major objectives of this initiative (Tongia et al., 2018). NREL has also been individually active in analyzing challenges in integrating renewable energy systems with the grid. Some of the solutions highlighted by them are advanced forecasting, reserve management, demand response and flexible generation sources, etc. (NREL, 2013). New metrics or improved metrics for better planning and integration was also highlighted but was rarely considered by researchers as a possible solution to challenges in grid integration. Particularly at an individual customer level, instrumentation cost for evaluating performance metrics is too high, and therefore, there is a huge knowledge gap between customer and utility. In addition to this, a misconception of zero maintenance in renewable systems like PV has further created ambiguity at the customer end. With a lack of instrumentation and the assumption of zero maintenance, the bridge between customers and utilities still needs to be strengthened.

3.1.1 LITERATURE REVIEW – GRID INTEGRATION ISSUES

Renewable energy integration poses challenges at different stages of the power system including customer and utility levels. Technically, grid integration limits the hosting capacity of renewable systems due to the necessity of maintaining permissible grid voltage and thermal rating of the grid (Matschoss et al., 2018). The following are some of the grid integration challenges concluded and mitigated in current literature.

- Power quality/stability/protection issues – Varun Kumar et al. (2016) presented power quality issues due to grid integration of renewable energy sources. This paper highlighted the role and importance of power electronic interface for efficient grid integration. Solutions to intermittency by installing smaller systems in a large geographical area than a large system at one place is also highlighted. Power quality issues like voltage swell/swag, real/reactive power problems, voltage stability, and harmonics have been presented with possible mitigation solutions through power electronics and FACTS devices.
- Generation uncertainty/variability issues – Few publications emphasized generation uncertainty issues in renewable energy while integrating with the grid. A detailed review of wind energy generation prediction and solutions to generation uncertainties has been presented by Ahmed et al. (2020). Large-scale energy storage systems were found to be a critical solution in

handling generation uncertainties. An extensive review of challenges and solutions in integrating small scale PV systems with the grid is presented by Alshaharani et al. (2019). Uncertainty in the generation is also reviewed with weather forecasting associated with ramping up and down conventional sources as a critical solution. Prusty and Jena (2018) presented a simulation model for assessing the risk associated with various severe events occurring in grid integrated PV using performing temperature-augmented probabilistic load flow (TPLF). The risk indices derived in this uncertainty modelling were found to be the solution in reducing risk in such systems.

- Control issues – Badal (2019) presented a detailed review of contemporary voltage control techniques used for the efficient operation of a microgrid with renewable energy sources. Ever-changing load poses a continuous challenge during microgrid operation, particularly in a standalone microgrid; different scenarios of this load dynamics have been simulated with control techniques and investigated for the performance of microgrid.
- Optimization issues – NREL (2013) presented a brief review of addressing methods to challenges faced due to uncertainty and variability in generation. Optimal solutions for addressing grid integration were found to be different for each scenario viz, advance forecasting, reserves management, demand response, and flexible generation. The cost-effectiveness of these methods will decide the best solution for addressing grid integration challenges.
- Maintenance issues – Bosman (2020) highlighted the need for a predictive maintenance tool for the benefit of PV owners after grid integration. This paper stresses the importance of monitoring solar radiance and temperature as two major parameters which, if deviated from rated values, can result in erroneous performance. Hence, the review concluded that predictive maintenance tools are required to ensure reduced interruptions in PV system after grid integration.
- Reactive power issues – Reactive power control has been a long term issue in grid-connected renewable systems, and this is generally a consequence of power quality issues in the grid. A detailed review of control techniques is presented by Sufyan et al. (2019) wherein three different techniques are discussed, and the results show that the combination of two control strategies simultaneously is effective than the individual technique. Hamrouni et al. (2019) presented a command-based technique of injecting reactive power into the grid by calculating inverter deliverable power based on their ratings and the respective power quality issues like voltage dips. This paper presents experimental results for a specific case of Symmetrical Grid Voltage Dips (SGVD) in the grid. Alenius et al. (2020) presents an autonomous reactive power compensation technique that measures grid impedance in real time. This technique was found to be better than conventional methods supported by experimental simulation on a grid-connected 3-Ph photovoltaic inverter. A detailed review of reactive power compensation techniques for renewable generation systems is presented by Sarkar et al. (2018). A variety of solutions including conventional devices like capacitor banks and static compensation devices like static synchronous compensators (SSC) etc. have

been reviewed. Response speed and control complexity are the two major parameters recommended to be considered while selecting a proper reactive power compensation technique.

Most of the research publications available are highlighting the issues at the power system level. Except a few, none has presented issues at customer/system owner/utility/installer levels. Hence, there is a huge gap in highlighting the issues at customer/utility levels. Addressing this gap, this chapter highlights the performance issues at the customer level and new metrics (as per standards) as a solution in bridging the gap between stakeholders after grid integration of PV systems. Performance analysis is a prerequisite for understanding the importance of new metrics. Performance analysis is a study that rates a system performance as per standards and gives insight into operational issues in a system and further helps in better future planning. Customers will be benefitted by following performance standards and protocols so that operational issues are minimized. Now that other issues have been already highlighted in the literature, performance issues and mitigations after grid integration of PV systems will be presented in this work. The chapter is organized as follows. Power system parameter issues specifically performance parameters relevant to both customer and utility sides, after grid integration, will be addressed in upcoming Section 3.2. Possible mitigation solutions are presented in Section 3.3. A summary of financial implications to customer/utility is explained in Section 3.4. A case study, as an example, demonstrating possible performance issues, mitigation, and financial implications relevant to customer/utility is presented in Section 3.5. Conclusions to this work are drawn in Section 3.6.

3.2 PERFORMANCE PARAMETER ISSUES TO CONSUMER AND UTILITY – PERFORMANCE ISSUES

The performance evaluation scheme of a grid integrated PV system gives an insight into how realistic the system is behaving with respect to the expected behavior. Like every learner/student is evaluated for his/her performance through standard test procedures, similarly, PV systems are also evaluated for their performance based on Standards IEC 61724. Various performance parameters of PV systems are measured first, and then some are derived as per standards. These parameters are indices that indicate the actual performance of the system relative to expected. Lower the actual, higher are the intermediate losses (issues) that need to be addressed. Usually, this practice is part of the daily/monthly/annual maintenance contract of the PV system but because of higher contract costs, particularly for individual consumers, performance evaluation is neglected causing operational, financial implications to both consumer and utility. This can result in operational issues like

- PV electrical breakdown (internal)
- PV mechanical breakdown (external)
- Inverter breakdown and wiring losses
- Deviation from actual contract or agreement (low voltage, lower power capacity, frequency deviations, etc.)

Renewable Energy Integration Issues

Technical standards IEC 61724: 1998 has been formulated and promoted as *Photovoltaic system performance monitoring – Guidelines for measurement, data exchange and analysis* (IEC, 1998). It describes the performance of the PV system at a systemic level as a whole instead of performance at the component level. The objective of these standards is to generate a comprehensive database from system performance gathered from PV systems across the globe so that comparison of systems could be done in terms of size, weather conditions, end-use, operation and maintenance, *etc.* This can also help an individual consumer in knowing the status of the installation. Usually, it happens so that after installation and commissioning, the consumer is bound to feed into the grid as per the expected output from the system subjected to an agreement with the utility. The consumer also expects monetary benefits like Feed-in Tariff (FiT) from the utility. However, performance issues like losses, improper maintenance, instrumentation errors, etc. may result in deviations in the actual output, in turn decreasing the financial benefits he/she expects. Hence, it is very much important for consumers to follow proper standards of performance. All the literature available reported the performance of simple grid-connected PV systems without any storage. This triggered a need for studying PV systems with battery storage for performance. Since the Dayalbagh Educational Institute (DEI) has a huge set-up of distributed solar photovoltaic systems with battery storage, performance parameter issues of PV-battery systems are presented in this chapter.

3.2.1 METHODOLOGY

3.2.1.1 Indian Standards

The Bureau of Indian Standards (BIS) adopted IEC 61724: 1998 "Photovoltaic system performance monitoring – Guidelines for measurement, data exchange and analysis" as Indian Standard IS/IEC 61724. This standard recommends the procedure for monitoring and assessing the performance of grid-connected PV systems.

3.2.1.2 Description of the SPV MicroGrid (SPVMG)

A brief description of the 40 kWp SPVMG considered for analysis in this work is given here in the form of specifications. The component specifications of the PV system are given in Tables 3.1–3.3. The structure of a solar photovoltaic microgrid is given in Figure 3.1.

3.2.1.3 Measured Parameters

Table 3.4 gives a list of parameters measured, as per IEC 61724, using appropriate sensors and devices. Irradiance is sensed by a pyranometer in the plane of the array measuring diffused radiation in W/m^2. Wind speed is an optional parameter. Considering occasional winds in the region, it was not included in the above parameters. Electrical parameters like voltage and current are measured for three different phases, and the total value is also measured. Although 231 parameters are measured and estimated by the measuring and data logging system, only those useful in this study are presented here.

TABLE 3.1
PV Specifications

S.No	Parameter	Rating/Value
1	Model	L24170 – Mono Crystalline Silicon
2	Make	Bharat Heavy Electricals Ltd. (BHEL), India
3	Open circuit voltage V_{OC}	42V
4	Short circuit current I_{SC}	5.2A
5	Voltage at max. power V_{mp}	35V
6	Current at max. power I_{mp}	4.86A
7	Maximum power P_{max}	170Wp
8	Max. system voltage	1,000V
9	Normally operating cell temperature (NOCT)	$45°C \pm 2°C$
10	Modules in each string	10
11	No. of strings	24
12	Nominal string voltage	240V

TABLE 3.2
Inverter Specifications

S.No	Parameter	Rating
1	Type	3-Phase
2	Output voltage	415V AC, 50Hz
	Input voltage	240–360V DC
3	Power rating	40 KVA
4	Efficiency	94% (at rated load)
5	THD	<4%

TABLE 3.3
Battery Specifications

S.No	Parameter	Value
1	Model	Lead acid
2	Make	EXIDE India
3	Ah capacity	400 Ah
4	Voltage per cell V_{pc}	2V
5	Charging rate	C 10
6	No. of batteries	120 (series)
7	DC system voltage	240
8	Depth of discharge	50% DoD

Renewable Energy Integration Issues

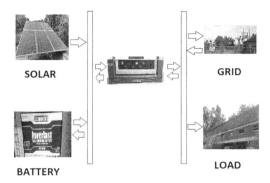

FIGURE 3.1 Structure of a solar photovoltaic microgrid.

3.2.1.4 Derived Parameters

Measured parameters presented in Table 3.4 are analyzed to derive energy and performance parameters, as per IEC 61724. These parameters are derived over different periods of monitoring *viz.* daily, weekly, monthly, and annually as shown in Table 3.5.

Mean daily Plane of Array (PoA) Irradiation values are calculated from measured irradiance values as follows:

$$H_{\text{daily}} = \int_{\text{SUNRISE}}^{\text{SUNSET}} G \; dt \left(\text{in kWh/m}^2 \text{/day} \right) \tag{3.1}$$

Energy quantities are calculated using respective power values over the recorded duration of time τ as follows:

$$\text{Net Energy}, E = \sum_{i=1}^{144} P_i * T \left(\text{in kWh} \right) \tag{3.2}$$

where time resolution (T) is 10 minutes (0.16 hours) with 144 such intervals in a day. Hence, 144 data logs are recorded every day by the data acquisition system, and every data-log contains time-weighted averaged sample values of all measured parameters. This time resolution can be varied by the user through HMI or software set-points of the data acquisition system. The lower the interval, the higher the resolution and vice-versa. Hence, every data log after a 10-min interval contains a time-weighted average of 1-min sampled values of each parameter.

Performance indices for such kind of systems are defined as yields, losses, and efficiencies. Yields are normalized energy quantities, losses are defined by the difference in yields between components, and efficiencies are the ratio of respective outputs to inputs. In addition to the above-derived parameters, the following indices are also calculated for a period of τ (normally a day – 24 h).

TABLE 3.4
List of Parameters Measured from the Solar Photovoltaic Microgrid

Parameter	Symbol	Unit
Plane of Array Irradiance	G	W/m²
Output voltage	V_A	V
Output current	I_A	A
Output Power	P_A	kW
Ambient Temperature	T_{amb}	°C
Cell voltage	V_{pc}	V
Battery voltage	V_{batt}	V
Battery current	I_{batt}	A
Battery power	P_{batt}	kW
Load voltage	V_{load}	V
Load current	I_{load}	A
Load power	P_{load}	kW
Grid voltage	V_{grid}	V
Grid current	I_{grid}	A
Grid power	P_{grid}	kW
Inverter		
Output voltage	V_{inv}	V
Output current	I_{inv}	A
Output power	P_{inv}	kW

Array yield (Y_A) is defined as the normalized value of array output (DC energy) w.r.t rated power of installed PV array

$$Y_A = E_A / P_{A,\text{rated}} \, \text{kWh/kW/day} \, (\text{or}) \, \text{h} \cdot \text{d}^{-1} \tag{3.3}$$

Final yield (Y_F) is defined as the normalized value of the final output of the system (AC energy) w.r.t rated power of installed PV array

$$Y_F = E_{ie} / P_{A,\text{rated}} \, \text{kWh/kW/day} \, (\text{or}) \, \text{h} \cdot \text{d}^{-1} \tag{3.4}$$

Reference yield (Y_R) is defined as the normalized value of irradiation* w.r.t the reference in-plane irradiance (G_{ref}). The universal value of G_{ref} is 1,000 W/m² as per Standard Test Conditions (STC). Effectively, it gives peak sun-hours for the period, normally a day.

$$Y_R = H_{daily} / G_{ref} \tag{3.5}$$

Renewable Energy Integration Issues

TABLE 3.5

List of Parameters Derived from the Measured Values of the Microgrid

Parameter	Symbol	Unit
Meteorology		
Plane of array irradiation (global) – daily	H_{daily}	kWh/m²/day
Cell temperature	T_{cell}	°C
Performance indices (d, m, A)*		
Array yield	Y_A	h/day
Final yield	Y_F	h/day
Reference yield	Y_R	h/day
Array capture losses	L_a	h/day
BoS losses	L_{BoS}	h/day
Performance ratio	PR	-
Capacity factor	CF	-
PV efficiency	η_{pv}	-
Overall system efficiency	η_{sys}	-
Savings in CO₂ emissions	CO₂	Kg/kWh

(or)

$$Y_R = \left(\int_0^\tau G \, d\tau \right) / G_{\text{ref}} \tag{3.6}$$

Losses are also normalized, in the sense that they are derived from the difference between yields. Losses, as indicated below, are given in terms of time (in hours) for which the array should operate at its rated capacity to cater for them.

$$\text{Array Capture losses}(L_A) = Y_R - Y_A \tag{3.7}$$

$$\text{BoS losses}(L_{\text{BoS}}) = Y_A x \ (1 - \eta_{\text{BOS}}) \tag{3.8}$$

Performance ratio (PR) defines the performance of the system taking into account the effect of all intermediate component losses right from the input radiation to the output AC energy produced. Hence, it is expressed as the ratio of final yield to the reference yield.

$$\text{PR} = Y_F / Y_R \tag{3.9}$$

$$\text{PR}_{dc} = Y_A / Y_R \tag{3.10}$$

$$\text{PR}_{ac} = Y_F / Y_R \tag{3.11}$$

58 Renewable Energy Integration to the Grid

Capacity Factor (CF) is a conventional index defined for any power generation system and is given as the fraction of the actual energy it could potentially produce if it were continuously operating at its rated capacity in a given time.

$$CF = (E_{ie})/(8760 * P_A)$$ (3.12)

Here E_{ie} is inverter export which is the net energy output of the system.

Efficiencies of various components of the system are defined and calculated as below.

$$PV\ Efficiency\ \eta_{pv} = E_A/(H_{daily} \times A_{pv})$$ (3.13)

A_{pv} is defined as the area covered by the PV array.

$$Overall\ system\ efficiency\ \eta_{sys} = \eta_{pv} * \eta_{load}$$ (3.14)

T_{amb} is measured using a temperature sensor, and T_{cell} is derived using the value of T_{amb} according to Gray et al. (2003)

$$T_{cell} = T_{amb} + \left[G \frac{(NOCT - 25)}{800} \right]$$ (3.15)

where NOCT is the nominal operating cell temperature and is taken as 45°C according to the manufacturer's datasheet for the PV modules installed in the PV systems considered in this work. CO_2 emissions, in the Indian scenario, from conventional power plants, are reported between 0.8 and 0.9 kg/kWh (Raghuvanshi et al., 2006). Also, a mean CO_2 emission factor of 0.82–0.85 kg/kWh is recorded by Central Electricity Authority (CEA), India (Bhawan & Puram, 2014). Hence, a CO_2 emission factor of 0.83 is used in this research work. In addition to this, other emissions are also calculated for SO_2, NO, and ash using emission factors of 0.00124, 0.00259, and 0.068, respectively, in kg/kWh (Agai et al., 2011).

3.3 PERFORMANCE RESULTS AND ISSUES AT THE CONSUMER END

This section analyzes the performance of the plant under study for 1 year. All the derived parameters mentioned in the previous section are comprehensively analyzed and depicted through plotting. R programming tool is used for computing and analysis of large sets of data logged for 1 year. The analysis is done, following the hierarchy of measured parameters and then derived parameters.

Solar irradiance is the basic input parameter measured in W/m^2. It is defined as power density or power received from the sun, per unit area of a receiver. Solar PV arrays are treated as receivers, and the incident solar radiation is measured in W/m^2 using a pyranometer, placed in the Plane of Array (PoA). Figure 3.2 depicts the daily irradiance values, as shown through box plots for each month. This parameter

Renewable Energy Integration Issues

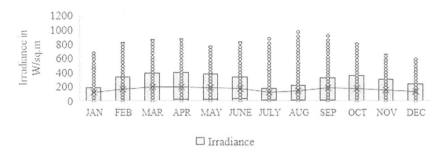

FIGURE 3.2 Box plot of monthly irradiance values, daily.

is measured throughout the day and night, and hence, the minimum value observed is always zero every day. However, the maximum irradiance received can be seen for each month. August has the highest received radiation value of 963 W/m². It was observed that August and September have the highest received radiation during the year. Irradiance is used to derive irradiation which is defined as energy received per unit area of the receiver in Wh/m². Since it is a time integral of irradiance value, irradiation or insolation is derived for a period *viz*. daily, weekly, monthly, and annually. Figure 3.3 clearly shows that the highest daily insolation value of 5,566Wh/m² was found to be in March followed by 5,158 Wh/m² in April. After these months, September, October, and May have higher insolation values. Hence, it is expected that these months are bound to be productive as far as PV array is concerned. However, this is not always true, in the sense that the output of the PV array depends on factors like temperature, tilt angle, wind speed, shade and dust, etc. other than irradiance.

Figure 3.4 depicts the daily averaged monthly load consumed and solar generation. The lowest consumption was noted in January and February along with December. Hence, grid export was also higher in these months.

As mentioned in the previous section, the considered PV system in this research work is analyzed as a solar photovoltaic microgrid and not as a simple grid-connected PV plant. PV is exclusively treated as a source, whereas grid and battery act as both source

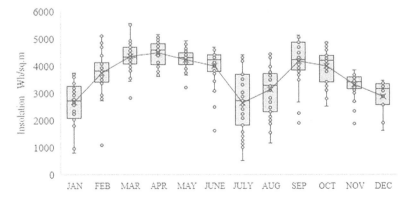

FIGURE 3.3 Monthly box plot of insolation in Wh/m²/day.

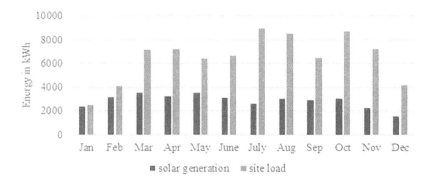

FIGURE 3.4 Average monthly distribution of solar generation and site load consumption (both in kWh).

and load, as per the mode of operation. Efficiencies of components and overall plant are shown in Figure 3.5. PV efficiency (η_{PV}) was higher during low-temperature months *viz.* January, February, and December. Overall plant efficiency (η_{sys}) nearly follows PV efficiency and is higher in the same months as PV and lower in the other months.

One of the first yields derived for the system is reference yield (Y_R). It is defined as the ratio of the mean plane of array insolation to the standard irradiance. It is a normalized quantity in hours (hrs). Taking standard irradiance as 1,000 W/m², Y_R gives the number of hours, and the standard irradiance input is available. It is represented in kWh/kWp or hours (hrs). Hence, this quantity is considered as input sun hours available to the total system, and it is expected that so many hours of output are available, which is practically not possible due to losses.

Figure 3.6 represents the average monthly Y_R based on the irradiance values measured. As expected, Y_R is minimum in the winter and rainy seasons due to climatic conditions. Winters in this region are as cold as 2°C with a lot of fog. The rainy season, although not a long and intense one, is anytime between July to September usually. According to Indian Meteorological Department (2016), more than half of annual

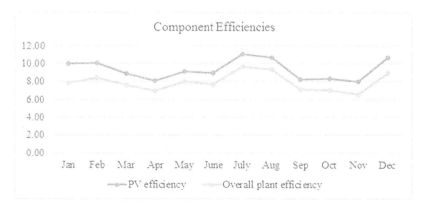

FIGURE 3.5 Average monthly system component efficiencies – PV, overall plant efficiency.

Renewable Energy Integration Issues 61

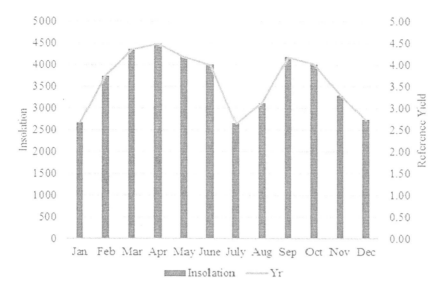

FIGURE 3.6 Daily average monthly reference yield with corresponding insolation levels.

rainfall was recorded in July which was also 54% more than the previous year. Hence, the lowest Y_R was observed in July. With Y_R as input to the PV array, the output from the array is derived as a normalized yield called array yield (Y_A). It is defined as the ratio of actual energy generated by the array to the rated capacity of the array. Hence, this is represented in kWh/kWp or hrs. However, the difference between the input yield Y_R and output yield Y_a is termed as array losses L_a. Figure 3.7 represents the yield and loss distribution for the considered PV array under study for 1 year. Array losses are reported to be more in hotter months than others due to higher temperatures despite higher reference yield. Since the temperatures in those months are much higher than the standard value of 25°C, there is a considerable amount of loss in those months. In addition to this, sandstorms are common in the region during summers increasing the soiling effect and reducing the output. Also during March/April and

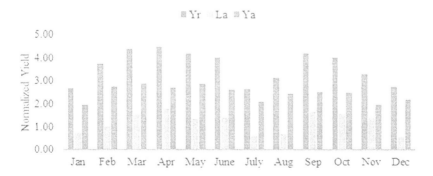

FIGURE 3.7 Distribution of PV array indices (yields).

September/October, huge dust is settled on the PV arrays due to seasonal harvesting in the neighborhood of the PV system. This adds to the total array losses during these months. Hence, winter and rainy months are found to be less troubled by array losses.

Total system indices precisely are Y_R, Y_A, and Y_F. The final yield is defined as the ratio of the total ac energy output of the system to the rated nameplate capacity of the system. It is defined in kWh/kWp or hrs, giving the number of sun hours available at the output of the system. Figure 3.8 depicts the distribution of yields throughout the year. Output yields along with intermediate losses are shown in Figure 3.9. L_a is the loss in the PV array, and L_{bos} is the loss in Balance of System (BoS) after the PV array. All the intermediate component losses, used for conversion, between PV array and output (load) viz. storage, inverter, and other wiring losses are summed up as L_{bos}. The performance ratio is a major performance index as per standards IEC 61724. It indicates the performance of any PV system taking into account all the intermediate losses. Hence, Figure 3.10 represents monthly PR along with corresponding losses. Higher losses result in lower PR as seen in March, April, September, October, and November. However, since PR is defined as the ratio of Y_F to Y_R, it is observed that lower radiation (Y_R) months have higher PR during winter and rainy months. The reason for this is the near STC temperature and minimum dust accumulation, in these months, which in turn reduces array losses (L_a) and improves PR.

PR can be defined for both the AC side and DC side. The ratio of DC output yield to input yield gives PR_{dc}, whereas the ratio of AC output energy yield to input yield gives PR_{ac}.

Figure 3.11 depicts monthly average values of PR_{dc} and PR_{ac}. Capacity factor is another major index shown in Figure 3.12. Its values range from 7% to 11%.

Irradiance and temperature are two major climatic factors affecting the performance of any PV system. The temperature is also dependent on irradiance. PV cell/module temperature is always greater than ambient mainly due to the current flow and partially due to the lower heat dissipation capability of encapsulated modules. The packing density of solar cells is another factor for increasing temperature. Hence, most of the incident irradiance is converted to heat rather than electricity. Figure 3.13 specifically plots cell temperatures between two-time slots viz. 7AM–6PM and 10AM–4PM. These are periods when irradiance is available, and hence, daily

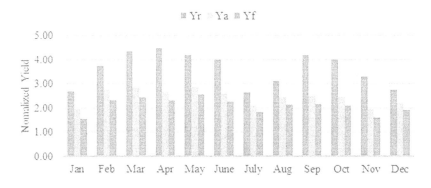

FIGURE 3.8 Distribution of total system indices (yields).

Renewable Energy Integration Issues

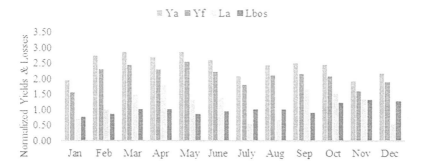

FIGURE 3.9 Distribution of output indices (yields) and losses.

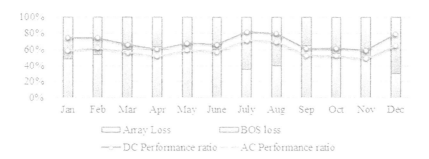

FIGURE 3.10 Comparative depiction of performance ratio (DC and AC) with corresponding losses.

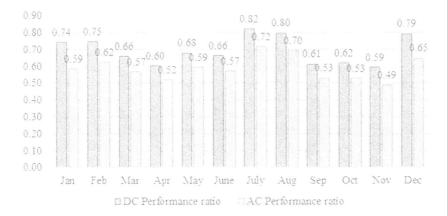

FIGURE 3.11 Annual distribution of major indices as per IEC 61724 Standard – PR_{dc} and PR_{ac}.

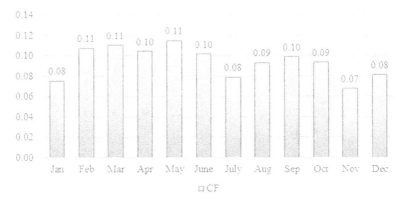

FIGURE 3.12 Annual distribution of capacity factor as per IEC 61724.

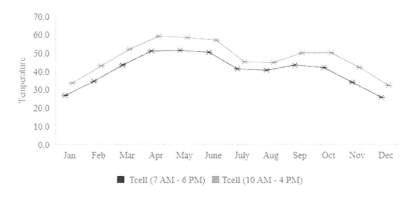

FIGURE 3.13 Monthly average variation in cell (module) temperature during specific time durations.

average values are reasonably appropriate. Higher values are observed evidently during March, April, May, and June. Daily average temperatures up to 60°C are observed in April and May. Figure 3.14 shows a clear indication of temperature depending on irradiance. Both the temperatures are plotted for ten different ranges of irradiance measured. Lower irradiance resulted in lower temperatures and vice versa. Specifically, variation in cell temperature is much more as the irradiance increases. The average cell temperature during the highest range between 900 and 1,000 W/m^2 is 63°C much greater than STC value. Figure 3.15 describes the reduction in PV efficiency due to an increase in temperature. The negative slope trend line shows the inverse relationship between the two parameters. Figure 3.16 describes the CO_2 annual savings for the considered period of analysis. Savings on CO_2 emissions and savings on SO_2, NO, and Ash are derived as 28,561, 39, 82, and 2,154 kg, respectively.

The following Table 3.6 is a consolidated comparative description of performance parameters at different locations across the globe hierarchically since the inception of standards to recent times. The list covers almost all regions such that any consumer can understand the local/regional performance issues due to weather, temperature, etc.

Renewable Energy Integration Issues 65

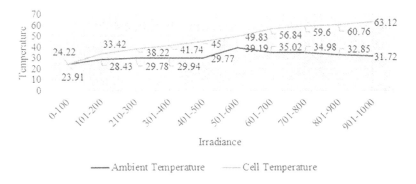

FIGURE 3.14 Variation in ambient and cell temperatures with respect to different irradiance ranges.

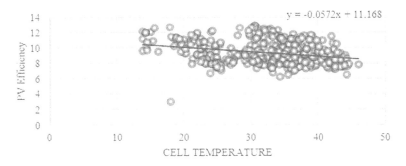

FIGURE 3.15 Inverse relation between PV efficiency and cell temperature.

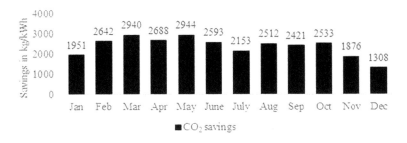

FIGURE 3.16 Monthly CO_2 savings in kg/ kWh.

3.4 MITIGATION OF ISSUES DUE TO GRID INTEGRATION

The following are some of the solutions to mitigate the power system parameter issues in the current literature

- IEEE 1547 Standards for Interconnecting Distributed Resources to Power system
- FACTS devices and controllers (Varun Kumar et al., 2016)

66 Renewable Energy Integration to the Grid

- Energy Storage
- Flexible generation and flexible demand (NREL, 2013)
- Frequency regulation (Ahmed et al., 2020)
- Regulatory policy (Tongia et al., 2018)
- Predictive maintenance (Bosman, 2020)
- Advanced/probabilistic forecasting (Prusty & Jena, 2019)
- Demand side management (Nwaigwe, 2019)
- Reactive power control (Alenius et al., 2020; Sarkar et al., 2018)

TABLE 3.6
Comparative Chart of Performance Indices Reported Across the World (Including This Work)

Research Study	System Size (kWp)	Energy Yield (kWh/kWp)	System Efficiency (%)	PR (%)	CF (%)	Location
Decker and Jahn (1997)	(1–5)	430–875	-	47.5–81	-	Germany
Lopez and Sidrach-de-Cardona (1998)	2	-	6.1–8.0	64.5	-	Spain
Moore et al. (2005)	4,900	1,324	-	60–70	-	Arizona
Mondol et al. (2006)	13	-	6.0–9.0	60–62	-	Ireland
Napat (2006)	5	1,278.4	-	71	-	Thailand
So et al. (2007)	3	1,170–1,378	7.9 – 9.0	63.3–75.1	12.0	Korea
Moore and Post (2008)	3,510	1,707	-	79	-	Tucson, Arizona
Kymakis et al. (2009)	171.3	715–1,850.5		67.36	15.26	Greece
Ayompe et al. (2011)	1.72	885.1	12.6	81.5	10.1	Ireland
Wittkopf et al. (2012)	142.5	1,138.8	11.2	81	15.7	Singapore
Eke and Demircan (2013)	2.73	1,412.5		72	23.2	Turkey
Sharma and Chandel (2013)	190	812.76	8.3	74	9.27	Khatkar-Kalan
Padmavathi and Daniel (2013)	3,000	1,372	-	72	15.69	Karnataka
Kumar and Sudhakar (2015)	10,000	1,579.8	-	76.20	17.68	Ramagundam
Adarmola and Vagnes (2015)	2.07	930.75	11.6	83.03	10.58	Norway
Farhoodnea et al. (2015)	3	1,387		77.28	15.7	Malaysia
Sundaram and Babu (2015)	5,000	1,755.65	5.08	85.5–92.3	-	Sivagangai
Pundir et al. (2016)	1,816	1,211	8.7	63.68	13.85	Roorkee
de Lima et al. (2017)	2.2	1,679	12.6	82.9	19.2	Fortaleza, Brazil
Sharma and Goel (2017)	11.2	1,339.5	12.5	78	15.27	Bhubaneshwar
Yadav and Bajpai (2018)	5	1,435.08	10.02	76.97	16.39	Lucknow
Current study	**40.8**	**793.5**	**8.5**	**59**	**9.0**	**Agra, India**

Renewable Energy Integration Issues

The PV generation system, in the case study, is a decade old 40 kWp rooftop installation with a 40kVA inverter and battery backup for 4 hours to cater for the full load of the building. This older version of the inverter has no integrated capability of reactive power compensation unlike the very recent versions of solar inverters. Hence, conventional automatic power factor control (APFC) with capacitor banks for reactive power compensation is still being used to control reactive power injected into the grid.

Most of the research publications available are highlighting the solutions at power system or grid level except for a few (Nwaigwe, 2019). Hence, there is a huge gap in highlighting the issues and their solutions at customer/utility levels. The upcoming section highlights the new metrics (as per standards) as a solution in bridging the gap between stakeholders after grid integration of PV systems. Performance analysis is a prerequisite for understanding the importance of new metrics which is already presented in the previous section.

3.5 FINANCIAL IMPLICATIONS TO CONSUMER/UTILITY

Results of a performance study of PV systems may allow consumers to understand the financial inputs required to be injected to maintain the expected output from the system. Periodic maintenance of the system is the best solution to mitigate financial losses for both consumer and utility. Now that in recent times the cost of grid injected power from renewables (PV) has come down to a large extent, the poor performance of any system may incur huge financial losses for both parties. Sometimes the instrumentation errors, as a result of poor maintenance, may also lead to heavy financial loss. For example, in the case of the same 40 kWp SPVMG, during 2017, malfunctioning of an instrument allowed continuous charging of battery throughout the day/night from the grid. This grid interaction of battery invited a heavy electricity bill form utility for the institute. Hence, a corrective action through the software set-points in the inverter was taken which reduced the grid interaction to 44% from 75% (Pritam Satsangi, 2019). This corrective action saved about 8,847 kWh (Rs 64,000) of energy for the institute for 6 months after correction compared to the same 6 month period in the previous year before correction. However, this kind of corrective action could be done only after continuous maintenance efforts and periodic monitoring. But this cannot happen with an individual consumer having a smaller-size PV system. This was an example of consumer losing financially due to grid integration and interaction. Similarly, the utility may expect or project a figure of feed-in power from consumers at a lower cost, but due to performance issues at the consumer end, the utility may have to end up purchase power from other market players at higher costs. Hence, both the parties are equally vulnerable, financially, to performance and maintenance issues in the PV systems after grid integration. One solution for mitigating adverse effects of grid integration on consumer and utility is to introduce new or improved metrics which gives a better picture of system condition accurately. Hence, the upcoming sections describe it as a solution with a case study on the same 40 kWp SPVMG installed in Dayalbagh Educational Institute.

3.6 CASE STUDY

There is always a scope of improvement of existing standards, by adding new performance indices, as reported by Trueblood et al. (2013) and Dierauf et al. (2013). World-renowned labs like National Renewable Energy Lab (NREL) and Sandia National Laboratories (SNL) have been active since the inception of the IEC standards and have been publishing various findings as technical reports. These reports are later accepted as revised/new standards of the IEC 61724 introduced in 2017. Since the scope of this research covers IEC 61724, this chapter throws light on the revised standards IEC 61724 – 1: 2017 (Edition 1). Although this revised version is released in three parts, part-1 is more relevant to the objective of this research work, and hence, it is studied and applied upon the microgrid under study. The published results in this work are novel and can contribute, their little, toward the stabilization of these standards by the end of the stability period. However, the results show that ambiguous situations between consumer and utility can be mitigated and help both parties.

3.6.1 Major Revisions in IEC 61724 (Standards)

The following technical changes were made to the previous version

- The new standard IEC 61724 part-1 defines the regulations to be followed in PV system monitoring by dividing the systems into different classes based on the accuracy of the measuring/monitoring equipment. In addition to this, overall performance evaluation based on monitored data is redefined.
- The other two parts IEC 61724 part-2 and part-3 defines performance evaluation in terms of power and energy respectively, based on monitored data.
- New temperature-corrected performance ratio and other performance indices are defined in addition to those given in the previous version.

The system under study here is the same SPVMG described in previous sections with dedicated load, battery storage, grid, AC loads, and power conditioning unit (PCU) – bi-directional Inverter (BDI)

3.6.2 New Performance Metrics

Some new performance metrics have been introduced in the revised version of the standards in addition to the ones defined in the previous version. This section describes each one of them in detail, and later, a consolidated table is presented with new and old metrics.

3.6.2.1 Soiling Ratio

It is a measure to quantify the effect of soiling on the output of a PV cell/module/array. Soiling ratio is defined as the ratio of the maximum power output of a soiled cell/module/array to the maximum power output of a cleaned one.

Soiling is a continuously effecting parameter on the output of the module. It is an accumulation of dust, bird droppings, and grime for a period until cleaning is done.

Normally cleaning of modules is done bi-annually (October & May). The months so-chosen are to allow dust accumulation mainly during harvesting periods in October and April. This is also reflected in the performance of the system presented in previous sections. Once the harvesting period is crossed, manual cleaning is taken up. However, rainfall in the geographical location of the system, Agra, is minimum (Indian Meteorological Department, 2016). This location falls under the category of rain-deficient (−20% to −59%) in winter, premonsoon, and monsoon, whereas it falls under the largely deficient (−60% to −99%) category for the postmonsoon period. Hence, dependency on rainfall for cleaning of PV systems is minimal and manual cleaning is preferred. As per new standards, IEC 61724: 2017 – part 1, the soiling ratio for the system under study is calculated using the method explained in the standards. Measurements of P_{max} (maximum power) and I_{sc} (Short circuit current) are taken for cleaned and uncleaned modules of the system under study for ten different strings after dust accumulation for 6 months. Since the period of dust accumulation falls between Oct 2017 and Apr 2018, the rainfall statistics in the location during the period shows zero rainfall in 4 out of 6 months with December and March receiving less than 10 mm unseasoned rainfall. Hence, considering the above, the soiling ratio was found to be between 0.75 and 0.96 on different days in ten strings as shown in Figure 3.17. The number of measurements taken is from 23 modules of ten different strings of the PV system. The measurements are taken during the mid-week of April after the harvesting period. The values of the soiling ratio 0.75–0.96 indicate the maximum power produced by the soiled module/array with respect to a cleaned one. Lesser the value, the more the effect of soiling. However, the last two measurements show values of 0.95 and 0.96 indicating a minimum soiling effect. This is due to rainfall, recent before the measurements, which naturally cleaned the module, and the measurements are taken two days later. Hence, the effect of soiling is not dominant in these two modules. Neglecting these two values, the other values clearly show up to 25% of output power reduction. An average soiling ratio of 0.83 is observed in the system of study which is also in agreement with Houssain et al. (2019) where 0.84 soiling ratio is reported. The trends in the soiling ratio shown by Micheli et al. (2017) indicate that variations in the soiling ratio are seasonal and site-specific depending upon activities nearby. In the present case, as reported earlier, March/April being the

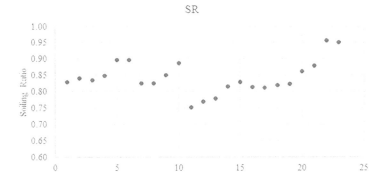

FIGURE 3.17 Soiling ratio of 23 modules from ten different strings.

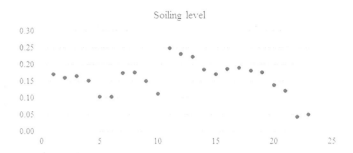

FIGURE 3.18 Soiling level of 23 modules from ten strings.

harvesting season, the soiling ratio is higher, and similar values are expected during October again due to harvesting in the nearby vicinity of the PV system.

3.6.2.2 Soiling Level

It is defined as the power loss caused due to soiling and is simply calculated as soiling level = 1 − SR

Figure 3.18 describes the soiling level values showing power deviations (losses) from expected power for each measurement with up to 25% deviation in some modules.

3.6.2.3 Performance Ratio

Although this parameter has been defined in the earlier version, there have been some additions to this metric, in new standards, in the form of some corrections due to weather parameters, particularly temperature.

3.6.2.4 Temperature-Corrected Performance Ratio

The performance ratio is a widely accepted metric in PV system operational guarantees and performance testing. It is a measure of the effectiveness in the conversion of radiation input to AC energy output against all possible intermediate losses including module reflection, soiling and shading, module mismatch, capture losses, BoS losses including inverter, battery and wiring losses, and failure losses during downtime. But performance database worldwide has reported, over the years, that PR is weather-dependent, particularly temperature. To elaborate, PR values are reported to be more optimistic in the winter season or for that matter colder regions due to lower and near-ideal cell temperatures during the period. Hence, PR might be more than the guaranteed value during the contract or agreement period between customer and utility/installer. However, the result is opposite during summer or hotter regions due to higher operating cell temperatures during the period, and hence, PR values are reported pessimistic or less than guaranteed. This seasonal variation in PR is site-specific and was never included during designing or operation, while it confuses both parties due to deviations, yearly observed. Hence, for the given annual PR value in the contract, it becomes necessary to eliminate the seasonal variations without actually changing the annual PR. This necessitates a site-specific correction in PR expression by including a local average cell temperature term, respective to location.

Renewable Energy Integration Issues

Hence, this section presents the computation of a temperature corrected PR as against uncorrected PR and showing the reduction in the seasonal variation by the proposed correction according to Dierauf et al. (2013) and IEC 61724-1 (2017). The following expressions describe the corrections made in PR

$$PR = \frac{Y_F}{Y_R} \tag{3.16}$$

$$Y_F = {E_{ie}} \Big/ {P_{A,\ rated}} \tag{3.17}$$

$$Y_R = {H_{daily}} \Big/ {G_{ref}} \tag{3.18}$$

where $H_{daily} = \int\limits_0^\tau G \ d\tau; E_{ie} = \int\limits_0^\tau P \ d\tau$ (τ is reporting time) – output energy

$$PR = {\int_0^\tau P \ d\tau} \Bigg/ {\dfrac{P_{A,\ rated} \times \int_0^\tau G \ d\tau}{G_{ref}}} \tag{3.19}$$

Now temperature-corrected PR can be calculated by including a correction term along with the rated power $P_{A,\ rated}$, termed as C_τ as shown below

$$\text{Temperature corrected } PR = {\int_0^\tau P \ d\tau} \Bigg/ {\dfrac{C\tau \times P_{A,\ rated} \times \int_0^\tau G \ d\tau}{G_{ref}}} \tag{3.20}$$

where

$$C\tau = 1 + \gamma \times \left(T_{mod,\tau} - T_{mod,avg}\right) \tag{3.21}$$

where $T_{mod,\tau}$ is the PV module temperature for the time period τ, and $T_{mod,\ avg}$ is the local average module temperature annually which was found to be 31.5°C for the site under study, for the considered year. This value may change for another year, if considered, depending on the weather data logs. Hence, it is important to find the accurate value of $T_{mod,avg}$ separately for each typical meteorological year (TMY) but satisfying the condition that annual uncorrected PR is the same as the annual temperature corrected PR (Dierauf et al., 2013; IEC 61724-1, 2017). γ is defined as the temperature coefficient for maximum power pertaining to the monocrystalline silicon module taken as $-0.45\%/^\circ C$ (Solanki, 2015).

3.6.2.5 STC-Corrected Performance Ratio

This PR is defined to compensate for the deviations due to the actual module temperature and STC temperature. To remind, Standard Test Conditions (STC) include reference values of radiation ($1,000\,W/m^2$), temperature ($25^\circ C$), and Air Mass Index – AMI (1.5). Like temperature-corrected PR defined in the previous section, STC-corrected PR is expressed as given below.

$$STC\ corrected\ PR = \frac{\int_0^\tau P\ d\tau}{\frac{C\tau' \times P_{A,\,rated} \times \int_0^\tau G\ d\tau}{G_{ref}}} \tag{3.22}$$

where

$$C\tau' = 1 + \gamma \times \left(T_{mod,\tau} - 25^\circ C\right) \tag{3.23}$$

3.6.2.6 Discussion and Results

Temperature-corrected performance ratio is described seasonally to depict the seasonal variation and its importance. Figure 3.19 shows the difference in uncorrected PR and corrected PR during winter months (January to mid-March). It is clear from the figure that higher values of uncorrected PR result during winter due to practically lower operating temperatures. This indeed is misleading the customer because these uncorrected PR portrays the system to be performing better, with higher PR, while it is not actually. However, this may not reflect in fulfilling the agreement between customer and utility. Even with higher values of uncorrected PR, the customer is not able to feed to grid/utility at maximum capacity. Hence, such systems need correction in evaluating PR with temperature correction. After applying the correction, the new corrected PR seems to be on the lower side, compensating for the temperature effects. As we go further into the spring season later, it is observed that the difference in both seems to be reduced gradually due to near-annual average temperatures.

The second part of the winter months (Nov and Dec) is shown in Figure 3.20 depicting uncorrected and corrected PR values. As discussed above, it was observed that gradually from November into December, the difference in both the values increases, and it continues into the next cycle of winter of the next year. Hence, it can be concluded that winters are prone to false high PR values misleading customer about the performance of the system. Hence, temperature-corrected PR is applied to

Renewable Energy Integration Issues

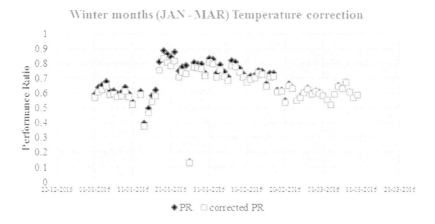

FIGURE 3.19 Comparison of uncorrected and corrected PR in winter months (JAN–MAR).

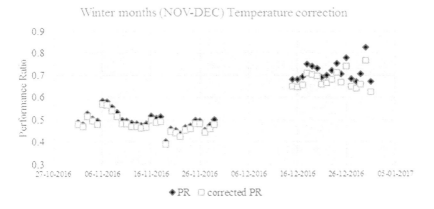

FIGURE 3.20 Comparison of uncorrected and corrected PR in winter months (NOV–DEC).

compensate for this effect. The gap shown in between during early December is due to the shut-down of the microgrid for maintenance.

Figure 3.21 shows variation in PR for typical summer months. It was observed that uncorrected PR values were pessimistic (on the lower side), misleading customers again regarding reduction in PR and indirectly creating risk for a contractor in view of deviating from the guaranteed performance. After temperature correction, it can be seen that PR is adjusted to its actual operating value, throughout the season.

Annual PR with temperature correction is depicted in Figure 3.22 where monthly values are shown. A clear difference between seasons is visible in PR. Months like March, September, and October were seen to be almost unaffected by temperature correction due to decent temperatures nearby ideal operating cell temperatures. An important condition to be satisfied while realizing temperature correction of PR is that the annual average uncorrected PR must be the same as the annual average

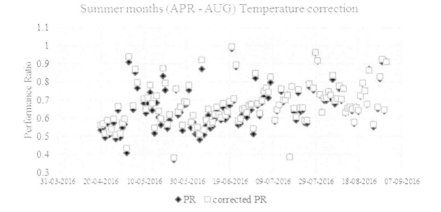

FIGURE 3.21 Comparison of uncorrected and corrected PR in the summer months (APR–AUG).

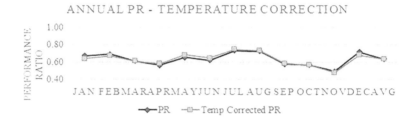

FIGURE 3.22 Annual temperature correction of PR.

corrected PR. Hence, this condition is shown to be satisfied in the same figure, depicting average values of both the same and overlapping each other.

After correction for local temperature issues, Figure 3.23 shows corrected PR for STC compared to uncorrected PR. This index, STC-corrected PR, is used to understand the deviations in PR due to deviations from standard temperature 25°C. Annual uncorrected PR compared with annual STC corrected PR are plotted.

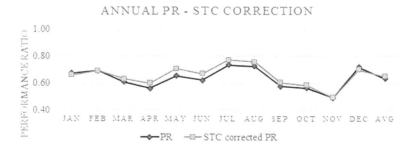

FIGURE 3.23 Annual STC correction of PR.

Renewable Energy Integration Issues

TABLE 3.7
Consolidated Performance Indices as per New Standards

S.No	Performance Indices	New Standards (IEC61724-1:2017)	Usage of the Index to Consumer/Utility
1	Performance ratio	63%	Indicates the overall performance of the system
2	Temperature corrected PR	63% (Annual)	Clears contract ambiguity in PR between consumer and utility
3	STC corrected PR	65.5% (Annual)	Alternative index supporting the above index
4	Soiling ratio	83% (6 months)	Both the soil indices indicate the loss of performance due to soiling and dusting.
5	Soiling level	17% (6 months)	Higher the level, higher is the performance/financial loss to the consumer

Summer months depicts STC correction more than winter months, which is in agreement with the standards according to the weather conditions of the microgrid under study. In this case, the average PR of both corrected and uncorrected are different with STC corrected PR being higher.

As a summary of the above discussion, Table 3.7 describes the consolidated performance indices as per new standards, their usage and relevance to both consumer and utility.

3.7 CONCLUSIONS

The role shift from a simple consumer to a prosumer (producer + consumer) is taking place at a faster pace nowadays. Schemes like Feed-in-Tariff (FiT), net metering, etc. have facilitated this shift. However, the role of a prosumer and challenges faced by him, as an important stakeholder, haven't been explored much specifically in grid-integrated renewable systems. Challenges faced at grid due to integration like power quality, reactive power control, stability, optimization, and protection have been already discussed by researchers. This chapter presents some of the issues due to PV system grid integration from both the consumer and utility perspectives.

Performance-based issues after grid integration, from the consumer perspective, are presented initially where the effect of component losses, weather, and temperature on the operational characteristics of a PV system is highlighted. The importance of proper awareness regarding performance issues is emphasized. Then methods of mitigating grid integration issues are listed with gaps in the available literature. Financial implications due to grid integration, specifically because of performance issues, are also highlighted. New/revised metrics as per standards IEC 61724: 2017 are shown to be a mitigating solution for performance parameter issues relevant to both consumer and utility. The results after the application of these new metrics for a PV microgrid, as a case study, shows that the ambiguity between consumer and utility can be mitigated and help both parties operationally and financially.

REFERENCES

Adarmola, M., & Vagnes, E. (2015). Preliminary assessment of a small scale rooftop PV-gird tied in Norwegian climatic conditions. *Energy conversion and Management, 90*, 458–465.

Agai, F., Caka, N., & Komoni, V. (2011). Design optimization and simulation of the photovoltaic systems on buildings in south east Europe. *International Journal of Advances in Engineering and Technology, 1*, 58–68.

Ahmed, S.D., et al. (2020, January). Grid energy challenges of wind energy: A review. *IEEE Access, 8*, 10857–10878.

Alenius, H., Luhtala, R., Messo, T., & Roinila, T. (2020). Autonomous reactive power support for smart photovoltaic inverter based on real-time grid-impedance measurements of a weak grid. *Electrical Power Systems Research, 182*, 1–14.

Alshaharani, A., Omer, S.A., Su, Y., & Mohamed, E. (2019). The technical challenges facing the integration of small scale and large scale PV Systems into the grid: A critical review. *Electronics, 8*(12):1443.

Ayompe, L.M., Duffy, A., McCormack, S.J., & Conlon, M. (2011). Measured Perforamance of 1.72kWp Rooftop Grid connected photovltaic system in Ireland. *Energy Conversion and Management, 52*, 816–825.

Badal, F. R. (2019). A survey on control issues in renewable energy integration and microgrid. *Protection and Control Modern Power Systems, 4*, 8.

Bhawan, S., & Puram, R. (2014). *CO_2 Baseline Database for the Indian Power Sector*. New Delhi: Central Electricity Authority, Ministry of Power, GoI.

Bosman, L. B.-S. (2020). PV system predictive maintenance: Challenges, current approaches, and opportunities. *Energies, 13*, 1398.

de Lima, C.L., de Araújo Ferreira, L., & de Lima Morais, F. H. B. (2017). Performance analysis of a grid connected photovoltaic system in North eastern Brazil. *Energy for Sustainable Development, 37*, 79–85.

Decker, B., & Jahn, U. (1997). Performance of 170 grid connected PV plants in North Germany - analysis of yields and optimization potentials. *Solar Energy, 59*, 127–133.

Dierauf, T., Growitz, A., Kurtz, S., Cruz, J.L.B., Riley, E., & Hansen, C. (2013). *Weather corrected Performance Ratio*. Golden, CO: National Renewable Energy Laboratory (NREL).

Eke, R., & Demircan, H. (2013). Performance analysis of a multi crystalline Si photovoltaic module under Mugla Climatic conditions in Turkey. *Energy Conversion and Management, 65*, 580–586.

Farhoodnea, M., Mohammed, A., Masou, T. K., & Elmenreich, W. (2015). Performance evaluation and characterization of a 3-kWp grid connected photovoltaic system based on tropical field experimental results: New results and comparative study. *Renewable and Sustainable Energy Reviews, 42*, 1047–1054.

Gray, J., Luque, A., & Hegedus, S. (2003). *Handbook of Photovoltaic Science and Engineering*. West Sussex, England: John Wiley & Sons.

Hamrouni, N., Younsi, S., & Jraidi, M. (2019). A flexible active and reactive power control strategy of a LV grid connected PV system. *Energy Procedia, 162*, 325–338.

Houssain, Z., Merrouni, A. A., Regragui, M., Bouachi, A., Hajjaj, C., Gennioui, A., & Ikken, B. (2019). Experimental investigation of the soiling effect on the performance of monocrystalline photovoltaic systems. *Energy Procedia, 157*, 1011–1021.

IEC. (1998). *61724: Photovoltaic System Performance Monitoring - Guidelines for Measurement, Data Exchange and Analysis*. Geneva: International Electrotechnical Commission.

IEC 61724-1. (2017). *Photovoltaic System Performance Part 1: Monitoring*. Geneva, Switzerland: International Electrotechnical commission.

Indian Meteorological Department. (2016). *Rainfall Statistics of India*. New Delhi: Indian Meteorological Department.

Kumar, B. S., & Sudhakar, K. (2015). Performance evaluation of 10 MWp grid connected solar photovotlaic power plant in India. *Energy Reports, 1*, 184–192.

Kymakis, E., Kalykakis, S., & Papazoglou, T. (2009). Performance analysis of a grid connected photovoltaic park on the island of Crete. *Energy Conversion and Management, 50*(3), 433–438.Lopez, M., & Sidrach-de-Cardona, M. (1998). Evaluation of a grid connected photovoltaic system in southern spain. *Renewable Energy, 15*, 527–530.

Kymakis, E., Kalykakis, S., & Papazoglou, T. (2009). Performance analysis of a grid connected photovoltaic park on the island of Crete. *Energy Conversion and Management, 50*, 433–438.

Matschoss, P., Bayer, B., Thomas, H., & Marian, A. (2018). The German incentive regulation and its practical impact on the grid integration of renewable energy systems. *Renewable Energy, 134*, 727–738.

Micheli, L., Muller, M., Deceglie, M. & Ruth, D. (2017). *Time Series Analysis of Photovoltaic Soiling Station Data*. Colorado, US: National Renewable Enery Lab.

Mondol, J., Yohanis, Y., Smyth, M., & Norton, B. (2006). Long term performance analysis of a grid connected Photovoltaic system in Northern Ireland. *Energy Conversion and Management, 47*, 2925–2947.

Moore, L., & Post, H. (2008). Five Years of operating experience at a large utility scale photovoltaic generating plant. *Progress in Photovoltaics: Research and Applications*, 249–259.

Moore, L., Post, H., Hayden, H., Canada, S., & Narang, D. (2005). Photovoltaic power plant experience at Arizona Public Service - A 5 Year Assessment. *Progress in Photovoltaics: Research and Applications, 13*, 353–363.

Napat, W. (2006). Two years performance of a 5kWp amorphous silicon GPV system. *21st European Solar Energy Conference*. Dresden, Germany.

NREL. (2013). *Integrating Variable Renewable Energy: Challenges and Solutions*. U.S. Department of Energy. NREL.

Nwaigwe, K. N. (2019). An overview of solar power (PV systems) integration into electricity grids. *Materials Science for Energy Technologies, 2*(3), 629–633.

Padmavathi, K., & Daniel, A. S. (2013). Performance Analysis of 3MWp grid connected solar photovltaic plant in India. *Energy for Sustainable Development, 17*, 615–625.

Pritam Satsangi, K. B. (2019). Real time performance of solar photovoltaic microgrid in India focusing on Self-consumption in institutional buildings. *Energy for Sustainable Development, 52*, 40–51.

Prusty, B.R., & Jena, D. (2018). An over-limit risk assessment of PV integrated power system using probabilistic load flow based on multi-time instant uncertainty modeling. *Renewable energy, 116, Part A*, 367–383.

Prusty, B. R., & Jena, D. (2019). A spatiotemporal probabilistic model-based temperature-augmented probabilistic load flow considering PV generations. *International transactions on Electrical Energy Systems, 29*(5), 1–13.

Pundir, K., Nandini, V., & Singh, G. (2016). Comparative study of perforamance of grid connected solar photvoltaic system in IIT Roorkee campus. *International Conference on Innovative Trends in Science, Engineering and Management*. New Delhi, INdia.

Raghuvanshi, S.P., Chandra, A., & Raghav, A. (2006). Carbon dioxideemissions from coal based power generation in India. *Energy Conversion and Management, 47*, 427–441.

Sarkar, N. I., Meegahapola, L.G., & Datta, M. (2018). Reactive power management in renewable rich power grids: A review of grid-codes, renewable generators, support devices, control strategies and optimization algorithms. *IEEE Access, 6*, 41458–41489.

Sharma, R., & Goel, S. (2017). Performance analysis of 11.2 kWp roof top grid connected PV system in Eastern India. *Energy Reports, 3*, 76–84.

Sharma, V., & Chandel, S. (2013). Performance analysis of a 190 kWp grid interactive solar photovoltaic power plant in India. *Energy, 55*, 476–485.

So, J. H., Young, Y., Yu, G., Choi, J., & Choi, J. (2007). Performance results and analysis of 3kW grid connected PV systems. *Renewable Energy, 32*, 1858–1872.

Solanki, C. (2015). *Solar Photovoltaics: Fundamentals, Technologies and Applications.* New Delhi: Prentice-Hall India Learning Pvt. Ltd.

Sufyan, M., Rahim, N., Eid, B., & Raihan, S.R.S. (2019). A comprehensive review of reactive power control strategies for three phase grid coonnected photovoltaic systems with low voltage ride through capability. *Journal of Renewable Sustainable Energy, 11*, 042701.

Sundaram, S., & Babu, J. S. (2015). Performance evaluation and validation of 5MWp grid connected photovltaic plant in South India. *Energy Conversion and Management, 100*, 429–439.

Tongia, R., Harish, S., & Walawalkar, R. (2018). *Integrating Renewable Energy into India's Grid: Harder Than It Looks.* New Delhi: Brookings India Impact Series.

Trueblood, C., Coley, S., Key, T., Rogers, L., Ellis, A., Hansen, C., & Philpot, E. (2013). PV Measures up for fleet duty. *IEEE Power & Energy, 11*, 33–44.

Varun Kumar, Pandey, A. S., & Sinha, S. K. (2016). Grid integration and power quality issues of wind and solar energy system: A review. *International Conference on Emerging Trends in Electrical Electronics and Sustainable Energy Systems* (pp. 71–80). IEEE.

Wittkopf, S., et al. (2012). Analytical performance monitoring of a 142.5kWp grid connected rooftop BIPV system in Singapore. *Renewable Energy, 47*, 130–138.

Yadav, S. K., & Bajpai, U. (2018). Performance Analysis of rooftop solar photovoltaic power plant in Northern India. *Energy for Sustainable Development, 43*, 130–138.

4 Voltage Issues in Power Networks with Renewable Power Generation

Sushil Kumar Gupta
DCRUST Murthal

Kapil Gandhi
KIET Group of Institutions

CONTENTS

4.1 Introduction 80
4.2 Voltage Collapse 80
4.3 VP and QV Curves 81
4.4 Voltage Drop in the Power Distribution Network 82
    4.4.1 Voltage Control through OLTC 83
4.5 Synchronous Generators as Ancillary Service Providers for Voltage Support 85
4.6 Issue of Voltage Stability and Voltage Collapse..88
4.7 RES Impact on Voltage Drop 90
4.8 RES Impact on Voltage Control 91
    4.8.1 RES Operation at Variable Voltage 91
    4.8.2 RES Operation at the Constant Voltage 92
4.9 Voltage Issue in Power Transmission Network with Renewable Energy Integration 92
4.10 Voltage Issue in Power Distribution Network with Renewable Energy Integration 93
    4.10.1 Voltage Issues with Solar PV...95
        4.10.1.1 Technical Challenges, Economical Benefits, and Environmental Impact of Solar-Grid Integration 96
    4.10.2 Voltage Issues with Wind DG..97
        4.10.2.1 Issue of Voltage Variation...97
        4.10.2.2 Voltage Dips Issue...98
        4.10.2.3 Harmonics...98
        4.10.2.4 Flickers...98

DOI: 10.1201/9781003271857-4

4.10.2.5 Wind Turbine Location 99
4.10.2.6 Low-Voltage Ride through (LVRT) Capability 100
4.11 Case Studies/Simulations 100
4.12 Summary 103
References 103

4.1 INTRODUCTION

The power distribution network must be operated within the acceptable range of the steady state voltage limits. The voltage can be adjusted by either controlling the reactive power flow or controlling the voltage directly. The on-load tap changer, voltage regulators, and shunt capacitors are used to control the voltage in the power distribution network. The power flow must be in a direction for proper operation of this equipment. The system voltage will be maximum near the generator and the minimum near the consumer side. An on-load tap changer (OLTC) transformer is the HV transformer with flexible tap settings. The OLTC is the necessary equipment of most of the substation power transformers. The reactive power demand of the load is compensated by the shunt capacitors which will raise the voltage. These capacitors can be installed in the power plants or in the substations. In large feeders, the shunt capacitor can't fulfill the demand of the reactive power, so the step voltage regulator is installed which works as an autotransformer [1].

The equipment of voltage and reactive power control required appropriate coordination. To sustain the voltages in the electric power distribution network within the permissible limits, the distribution network operators (DNOs) must operate these equipment through conventional controllers. Many researchers proposed the automatic distribution system to resolve the problem of voltage control with real-time control because the off-line control has dependency on the communication channels [2,3]. The automatic distribution system with real-time control is a costly affair.

The impact of the renewable energy penetration on the performance of the distribution network with solar and wind integration is discussed. The challenges and benefits of solar PV are explained in detail. The problems with the voltage control when the wind turbine penetration is high are also discussed. A case study using a 13-bus hybrid microgrid which consists of solar PV, wind distributed generators (DG), and battery bank is discussed with the help of model and simulation.

4.2 VOLTAGE COLLAPSE

According to the collaborative work of IEEE and CIGRE on stability terms, the voltage collapse in the power system can be defined as *the process by which the arrangement of events affects voltage instability and leads to a system failure or undesirable low-voltage profile in a significant area of the power system.* Generally, voltage collapse occurs due to voltage instability or voltage transients. The major reasons of blackouts throughout the world are voltage collapse and voltage instability (e.g. New York 1970, France 1978). The severity of voltage collapses has prompted to research efforts on voltage stability and voltage collapse. There are numbers of static and dynamic analysis techniques already planned to identify the effect of voltage instability or issues. Various researchers proposed multiple indices (P index, L index etc.) to identify the effect of point of voltage collapse on the system. Some indices are very useful for planning and designing of power system, and some are used to monitor the condition of the system stability [4].

4.3 VP AND QV CURVES

The relationship between the active power injection in the power system and the corresponding change in bus voltage shows through the VP curve which is also known as 'nose curve'. Figure 4.1 represents the VP curve. The stable and operating region of the power system is in the upper part of the curve, while the unstable region of the power system is in the lower part. The bus voltage depends on the active power injection, reactance of that line, and the power factor of the load. The bus voltage decreases with the increase in the system load and beyond the permissible limits; it will cause the instability in the system. The stability limit of the system is represented through the tip of the 'nose curve'. The power flow must be continuous to attain the nose curve of the power system. The Jacobian matrix for the equations of power flow becomes singular. The regular power flow's solution does not converge. The load flow equations of the power system need to restructure to overcome from the stability problem due to continuous power flow at all possible loading conditions.

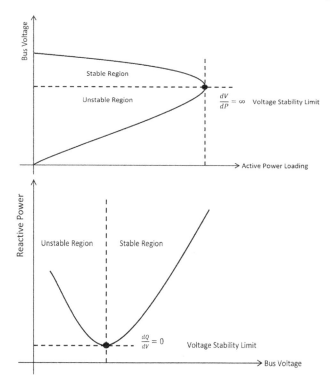

FIGURE 4.1 VP and QV curves of the power system.

It will allow solving the load flow problem for the stable and unstable loading conditions [1,2,5,6].

Another important characteristic for analysis of the voltage stability is the QV curve of the system. These curves represent the variation of bus voltages with respect to the reactive power injections in the system. Figure 4.1 represents the QV curve. The QV curve is used by many utilities due to its advantages on convergence and direct relation with reactive power compensation. The voltage stability limit is represented by the bottom area of the curve where $\Delta Q/\Delta V$ is equal to zero. The right side of the QV curve represents the operating region and is more stable because the reactive power is directly related to the bus voltage in this area. The left side of the QV curve is unstable because the reactive power is inversely related to the bus voltage in this area [6,7].

The most conventional methods to determine the active power and reactive power margin is VP and QV curves. But the main challenge with this method is that the large number of curves for various operating regions is required to determine the exact condition of voltage stability. Each curve requires a large number of power flow data to generate. It will take lots of time to execute, so it is not feasible for online monitoring of voltage stability for the power system consisting a large number of buses [7,8].

4.4 VOLTAGE DROP IN THE POWER DISTRIBUTION NETWORK

The single-line diagram from grid to load representing the voltage across different terminals of the system is shown in Figure 4.2. The current in the feeder is the variable of the complex apparent power across the load $S = P_L + jQ_L$ and the terminal voltage across the load V will be

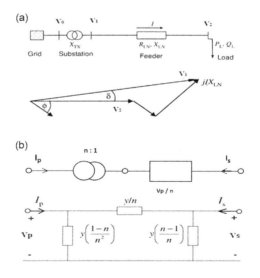

FIGURE 4.2 (a) Single-line diagram and phasor diagram of the power distribution network. (b) Load tap changer representation.

Voltage Issues in Power Networks

$$I = \frac{S}{V} = \frac{P_L - jQ_L}{V} \tag{4.1}$$

The voltage drop across the feeder of the system is given as

$$\left| \underline{V_1} - \underline{V_2} \right| = \underline{I} \left| (R_{LN} + jX_{LN}) \right|$$

$$= \left| \frac{(R_{LN}P_L + X_{LN}Q_L) - j(X_{LN}P_L - R_{LN}Q_L)}{\underline{V_2}} \right| \tag{4.2}$$

For the low power flow, the voltage angle (δ) between receiving end voltage V_2 and sending end voltage V_1 in Equation (4.2) is small, and the voltage drop $\Delta V = \left| \underline{V_1} - \underline{V_2} \right|$ can be determined by

$$\Delta V = \frac{R_{LN}P_L + X_{LN}Q_L}{U_2} \tag{4.3}$$

It can be determined from Eqs. (4.2) and (4.3) that the voltage drop will increase with the system load, and the overall voltage profile of the systems will decrease near the end terminal. The voltage regulation (percentage difference between voltage at minimum and maximum load) occurs due to the voltage drops in the distribution network. OLTC is used to adjust the output voltage of the transformer by changing the voltage/tap ratio, and the reactive power demand on the feeder can also be balanced using capacitor bank/panels [8–10].

4.4.1 VOLTAGE CONTROL THROUGH OLTC

The voltage ratio of the transformer can be easily changed by changing the numbers of turns in primary windings or secondary windings with load tap changer (LTC). The symbolic representation of the transformer with an LTC feature and its equivalent circuit diagram is shown in Figure 4.2. The LTC can be located at any side, i.e., primary terminal side or the secondary terminal side of the transformer. Notations V, I, η, and y in Figure 4.2 represent the voltage, current, transformer turns-ratio, and transformer admittance, respectively [10,11].

There are two types of LTC: off LTC in which the turns-ratio can be varied only when the connected transformer is disconnected from the supply mains or at No-Load, and OLTC where the tap setting position can be changed also when the power transformer is connected with the supply mains or carrying the load. The basic connection of OLTC is represented in Figure 4.3a. The OLTC controller regulates the substation secondary bus voltage V_1 constant within the permissible range

$$V_{LL} < V_1 < V_{UL} \tag{4.4}$$

where

$$V_{LR} = V_{set} - 0.5 V_{DB} \tag{4.5}$$

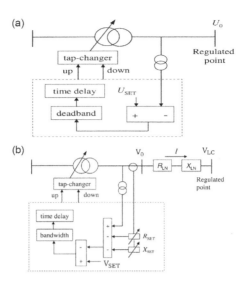

FIGURE 4.3 (a) Basic OLTC arrangement. (b) OLTC with LDC feature.

is the lower range of the voltage

$$V_{UR} = V_{set} + 0.5 V_{DB} \qquad (4.6)$$

is the upper range of the voltage. V_{set} is the set point voltage, and V_{DB} is the dead band.

The line drop compensation is activated with OLTC arrangement as shown in Figure 4.3b. The function of compensation functions to keep the voltage constant at the remote bus of the power system. The voltage drop in the system depends on the parameters (line current and line impedance) of the system which can be evaluated using line drop compensation. The voltage corrections are within the range when load is approx. constant [11,12].

$$V_{LR} < V_{LC} < V_{UR} \qquad (4.7)$$

Appropriately adjusting line resistance R_{LN} and line reactance X_{LN} to the voltage or turns ratios of the CT and PT is given by

$$R_{set} = \frac{N_{CT}}{N_{PT}} * R_{LN} \qquad (4.8)$$

$$X_{set} = \frac{N_{CT}}{N_{PT}} * X_{LN} \qquad (4.9)$$

where R_{set} and X_{set} are line drop compensation setting range for the resistive and reactive compensation of the transformer.

Voltage Issues in Power Networks

N_{CT} represents the voltage ratio of the current transformer, and N_{PT} represents the turns voltage of the potential transformer.

The voltage of system at the load compensation during minimum and maximum load powers is determined as

$$V_{LC} = V_{0,\,max} - I_{max} \left(R_{LN} \cos\varnothing + X_{LN} \sin\varnothing \right) \qquad (4.10)$$

$$V_{LC} = V_{0J\,min} - I_{min} \left(R_{LN} \cos\varnothing + X_{LN} \sin\varnothing \right) \qquad (4.11)$$

where

$V_{0,\,max}$ and $V_{0,\,min}$ represent maximum and minimum values of voltage of sending end.

I_{max} and I_{min} represent maximum and minimum values of line current of the system.

$\cos\varnothing$ is the pf of the appropriate transformer.

Generally, LDC function is disabled in OLTC operated transformers to make the system control easier and to minimize the errors. Due to this the load parameters are in control of its LDC as shown by Equations (4.10) and (4.11). The variation in the power factor between the load center and OLTC will affect the overall performance of the LDC. The performance can be decreased if the X / R ratio of the transformer is not appropriately attuned in that system.

The load side voltage is monitored through a voltage relay. If the voltage will cross the deadband which has a range of 2 volt generally, a relay is energized to operate the tap mechanism of OLTC until the voltage again shifted in the band or until the maximum or minimum tap is reached. The time to reset the voltage is between 10 and 100 seconds. In a large transformer, it can reach up to 2 minutes maximum. The range of power transformer is -10% to $+10\%$ in bulk power transformer in the 32 steps of 0.625% each [12–14].

4.5 SYNCHRONOUS GENERATORS AS ANCILLARY SERVICE PROVIDERS FOR VOLTAGE SUPPORT

The generation of reactive power on demand is the best parameter to evaluate the performance of the any generator and the synchronous generators responses immediately to support reactive power. Figure 4.4 shows the limits on active and reactive production for a typical generator (capability curve of the generator). The output of the generator is limited by its current carrying capability. The current carrying capability represents the MVA limit for the generator at rated voltage. The limit is determining in MVA, not in MW due to the varying power factor of the load. The magnetic field will increase to raise the generator's terminal voltage to produce the necessary reactive power. This increasing magnetic field will further increase the field current of the rotor. After the particular limit, the field winding starts to heat due to their limited capacity. The synchronizing torque is reduced when less reactive power is absorbed due to limited magnetic flux pattern of the stator, which can affect the chance of losing synchronism with the system.

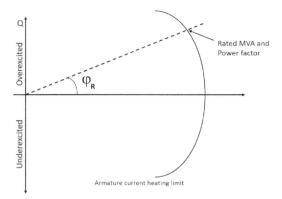

FIGURE 4.4 P-Q limitation due to armature current.

The complex power output from the synchronous generator is given by

$$S = P + jQ \tag{4.12}$$

$$S = V_t I_a^* \tag{4.13}$$

$$S = V_t I_a (\cos\varnothing + j \sin\varnothing) \tag{4.14}$$

$$P^2 + Q^2 = (V_t I_a)^2 \tag{4.15}$$

This equation is of circle with center at (0,0) and radius equal to $V_t I_a$ (MVA rating) as shown in Figure 4.4.

The generator generates the active and reactive powers as per load demand under normal operating conditions. If the active power demand is not balanced by active power supply, its frequency shall be affected and if reactive power demand is not balanced by reactive power supplied by the generator, its voltage shall be affected [15,16].

Voltage deviation of the generator is the function of both active and reactive powers and is given by the following equation:

$$|\Delta V| = \frac{R \cdot P + X \cdot Q}{|V_R|} \tag{4.16}$$

where $|V_R|$ is the magnitude of receiving terminal voltage and P and Q represent the active power and reactive powers, respectively, at receiving terminal.

When the resistance of power system network is low,

$$|\Delta V| = \frac{X \cdot Q}{|V_R|} \tag{4.17}$$

$$Q = f(V) \tag{4.18}$$

Voltage Issues in Power Networks

$$P = \frac{|E||V|}{X} \quad (4.19)$$

The reactive power production depends on the field excitation of the generators, so the excitation must be continuously controlled to match the demand; otherwise, the voltage at the bus terminals may go beyond the prescribed limits due to reactive power mismatch. The field excitation system consists of automatic voltage regulator (AVR) and an exciter to provide the required field current to the rotor. The work of the voltage regulator is to control the excitation through exciter to make constant terminal voltage irrespective of reactive power demand at any load condition.

Let us observe the effect of the varying excitation on generator for constant output power P_G and terminal voltage (V). The synchronous generator power equation is given by

$$P_G = |V||I|\cos\varnothing = \frac{|E||V|}{X}\sin\delta \quad (4.20)$$

For constant P_G and $|V||I|\cos\varnothing$,

$$E_f = V + jI_a X_s \quad (4.21)$$

The single-line diagram and phasor diagram of the synchronous generator shall be as shown in Figure 4.5. The center locus of I_a vector shall be on line ab since $|I|\cos\varphi$ on this line shall be constant. Similarly, E_f vector shall be on horizontal cd line as $E_f\sin\varphi$ on this line shall be constant. If I_a is equal to I_{a1}, i.e., the excitation requirement shall be E_{f1} called normal excitation. And if I_a is equal to I_{a2}, i.e., at leading power factor, excitation requirement shall be equal to E_{f2}. The above discussion is applicable for lagging power factor load because more than 90% load is inductive, it requires under excitation, and to supply the leading power factor, synchronous

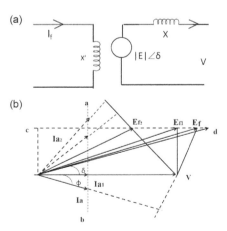

FIGURE 4.5 (a) Representation of synchronous generator. (b) Phasor diagram of synchronous generator.

generator requires leading excitation. Therefore, the reactive power flow controls by the excitation system of the generator to maintain constant output voltage [17,18].

The capital cost and running cost with extra equipment (e.g., voltage regulator, an exciter) are considerable amount to charge for the voltage control function. The voltage control cost depends on these equipment in vertical power system. But when the microgrid introduces, the additional reactive power demand can be fulfilled by decreasing the active power production which increases the total running cost of voltage control of the system. The prices are not linear with the load in nature (or may be exponentially) because the cost of voltage control may be very high after the prescribed limits.

4.6 ISSUE OF VOLTAGE STABILITY AND VOLTAGE COLLAPSE

The voltage stability of the system is defined as he capability of the power system to preserve acceptable voltage range at all buses in the power system under normal conditions after being subjected to a disturbance. Voltage instability arises due to nonfulfilment of the reactive power demand of the system.

Let a generator of constant voltage V_s is supplying inductive load through a very short transmission line whose impedance is Z, as shown in below Figure 4.6.

$$V_S = V_R + IX \tag{4.22}$$

Here $V_R I = Q$ (Q is reactive power demand)

$$V_S = V_R + X\left(\frac{Q}{VR}\right) \tag{4.23}$$

$$V_R^2 - V_R V_S + QX = 0 \tag{4.24}$$

$$V_R = \frac{V_S \pm \sqrt{V_S^2 - 4QS}}{2} \tag{4.25}$$

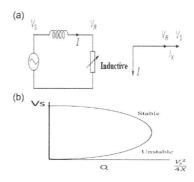

FIGURE 4.6 (a) Transmission line representation. (b) Stable and unstable regions of the system.

Voltage Issues in Power Networks

$$V_R = \frac{V_S}{2} \pm \frac{\sqrt{V_S^2 - 4QS}}{2} \qquad (4.26)$$

Receiving end voltage V_R must vary with the reactive power demand as shown below (Figure 4.6):

For maximum reactive power demand of the load, $Q_{max} = \dfrac{V_S^2}{4R}$, $V_R = \dfrac{V_S}{2}$

When the reactive power demand from the generator is beyond Q_{max}, system shall be in unstable conditions. The continuous disturbance in the system lead to the voltage instability which further becomes the major reason of the voltage collapse which can create panic for system security, stability and reliability. The demand of the reactive power is different from bus to bus in the system which can't fulfill by the generator in the occurrence of voltage collapse. During voltage collapse, the voltage starts to decrease continuously up to the critical value of the operating parameters of switchgear protection. The voltage angle and frequency of the system may constant or may be vary with the system voltage. It will occur due to three phase fault or due to inadequate reactive power support at certain critical network buses of the system.

From Equation (4.22),

$$V_S I = IV_R + I^2 X \qquad (4.27)$$

$$Q_S = Q_R + I^2 X \qquad (4.28)$$

$$Q_S = Q_R + \frac{Q_S^2}{V_S^2} \qquad (4.29)$$

$$Q_S = Q_R + \frac{Q_S^2}{4Q_{R\,max}} \qquad (4.30)$$

$$Q_S^2 - 4Q_{R\,max}Q_S + 4Q_{R\,max}Q_R = 0 \qquad (4.31)$$

$$Q_S = \frac{+4Q_{R\,max} \pm \sqrt{16Q_{R\,max}^2 - 16Q_{R\,max}Q_R}}{2} \qquad (4.32)$$

$$Q_S = 2\left[Q_{R\,max} \pm \sqrt{Q_{R\,max}^2 - Q_{R\,max}Q_R}\right] \qquad (4.33)$$

Differentiate w.r.t to Q_s in Equation 4.29:

$$1 = \frac{dQ_R}{dQ_s} + 2Q_s \frac{X}{V_S^2}$$

$$\frac{dQ_R}{dQ_s} = 1 - 2Q_s \frac{X}{V_S^2}$$

$$\frac{dQ_R}{dQ_s} = 1 - \frac{2Q_s}{4Q_{R\,max}}$$

$$\frac{dQ_R}{dQ_s} = 1 - \frac{2\left[Q_{R\,max} \pm \sqrt{Q_{R\,max}^2 - Q_{R\,max}Q_R}\right]}{2Q_{R\,max}}$$

$$\frac{dQ_R}{dQ_s} = \sqrt{1 - \frac{Q_R}{Q_{R\,max}}} \tag{4.34}$$

$$\frac{dQ_s}{dQ_R} = \frac{1}{\sqrt{1 - \dfrac{Q_R}{Q_{R\,max}}}} = \text{Voltage collapse Proximate indicator}$$

Thus, when the Q_R increases, voltage collapse proximity indicator (VCPI) increases. And when Q_R increases to $Q_{R\,max}$, VCPI becomes infinity. VCPI is the very sensitive and an important indicator to determine the risk of voltage collapse [19,20].

4.7 RES IMPACT ON VOLTAGE DROP

Renewable energy sources (RES) or DG can be integrated with the main grid straight using generators or through the power electronic converters. The RES deals with the dc power and does not require reactive power, so they don't exchange reactive power while RES works on AC, generates, or absorbs reactive power within the distribution system. When the RES is working at a certain system voltage by regulating the reactive output power can generate reactive power. Induction generator-based RES always absorbs reactive power.

For the system consisting of RES and load represented in Figure 4.7, the voltage drops on particular feeder of the power system can be determined by

$$\Delta V = \frac{V_1 - V_2}{V_2} = \frac{R_{LN}(P_L - P_{DG})_1 + X_{LN}(Q_L - (\pm Q_{DG}))}{V_2} \tag{4.35}$$

It represents that if the RES does not participate in the exchange of the reactive power, then it will affect the voltage drop across the system feeder. In RES, power generated is unpredicted, and sometimes, the generated power is higher than the load on that particular feeder. Due to this excess generation, the power will start

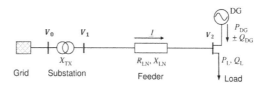

FIGURE 4.7 Voltage drops in the power distribution network with RES.

to flow to the substation and cause voltage rise at the bus. If RES can also absorb the excess reactive power insertion, it will change the voltage rise at the bus (Equation 4.35). The voltage rise depends on RES power (active and reactive) relation to the load power (active and reactive) and the X/R ratio of the transmission line [20–22].

4.8 RES IMPACT ON VOLTAGE CONTROL

When the operating voltage of the distribution system is near to the maximum permitted voltage V_{max} then the losses in traditional distribution network can be minimized at constant power and it is obtained by changing the terminal setting of OLTC upper limit V_{UL} and the feeder capacitor switching-off voltage V_{OFF} near to V_{max}. The analysis of the impact of RES penetration in the distribution system on the voltage control is based on permitted limits of voltage [22–24].

4.8.1 RES OPERATION AT VARIABLE VOLTAGE

The single-line diagram is shown in Figure 4.8 with RES in the feeder of the system. The terminal voltage can't be controlled by the RES. When the voltage rise occurs due to RES in the distribution system, it may cause over voltage between two buses. It can happen when the bus voltage will increase to V_{max} when the capacitor, installed at feeder, is energized. Therefore, the overvoltage at RES terminal buses can be prevented by decreasing the voltage setting of feeder capacitor.

The variation of the voltage is investigated by presuming that the load in the Figure 4.8 increases from P_{L1}, Q_{L1} to P_{L2}, Q_{L2}, through the constant RES output power. Therefore, due to the increasing load, the voltage at load bus of the system will reduce from V_{21} to V_{22}, which can be determined by

$$V_{21} - V_{22} = \frac{R_{LN}(P_{L2} - P_G) + X_{LN}(Q_{L2} - (\pm Q_G))}{V_{22}} \\ - \frac{R_{LN}(P_{L1} - P_G) + X_{LN}(Q_{L1} - (\pm Q_G))}{V_{21}} \quad (4.36)$$

Equation (4.36) represents that the RES doesn't participate in exchange of the reactive power or generates reactive power within the distribution system, and the voltage varies with the system load in the presence of RES. When the RES starts to

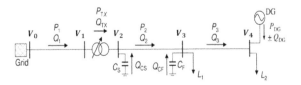

FIGURE 4.8 RES effect on voltage control.

absorb the reactive power, its impact on the system will be subject to active power and reactive power generation with respect to the load power and the X/R ratio of the line [24,25].

4.8.2 RES OPERATION AT THE CONSTANT VOLTAGE

The voltage of the distribution system in the presence of RES can be controlled by varying the reactive power (Q) output in the rotor field excitation system of the generator. The response of the excitation system (t_{ES}) is much faster than the response of the capacitors (t_{FC}, t_{SC}) and the response of the OLTC (t_{OLTC})settings. Hence, the time delay for RES excitation is depend on OLTC and capacitors which are related as

$$t_{ES} < t_{FC} < t_{SC} < t_{OLTC} \tag{4.37}$$

The terminal voltage of RES can be controlled by RES until the voltage is in the range of the reactive power limits which is specified by

$$Q_{RES} \mid_{min} \leq Q_{RES} \leq Q_{RES} \mid_{max} \tag{4.38}$$

The single-line diagram is shown in Figure 4.8 where V_4 remains constant by regulating the output reactive power of RES. The voltage of the distribution system depends on the load, the voltage will change when the load changes, and this change can be smaller when the V_4 also varies with the load. When the voltage variation is low on the feeder bus, the OLTC setting terminal voltage also needs to decrease to avoid unnecessary tripping of feeder capacitor. Therefore, the OLTC terminal voltage can be improved to function the power distribution system even closer to its peak permitted voltage and to reduce the losses [20–22].

4.9 VOLTAGE ISSUE IN POWER TRANSMISSION NETWORK WITH RENEWABLE ENERGY INTEGRATION

The high penetration of renewable energy integration (solar and wind) in the grid through the power electronic converters and inverters disturbs the stability of the grid. The power electronic converters and controllers are very fast to respond, but the grid response time is higher due to old technologies. It challenges the grid stability and reliability. The power generation through renewables completely depends on the weather conditions. It is difficult to predict the exact generation with accuracy. The unpredicted power integration in transmission network generates the harmonics and voltage spikes.

On 30 July 2012, the northern grid collapsed due to unpredict demand of electric power by some states of India due to increasing load. The permitted range of frequency in India is 49.5–50.25 Hz. Due to increasing power demand, frequency will slightly disturb which will further collapse the grid. The northern grid shut down first and then the free flow of excess power in remaining four grids also collapsed the eastern grid and northern-east grid. It is the largest blackout in the India as well as in the world. More than 600 million people were affected though this blackout (Figure 4.9). Electric power was restored in 36–48 hours [25–27].

Voltage Issues in Power Networks

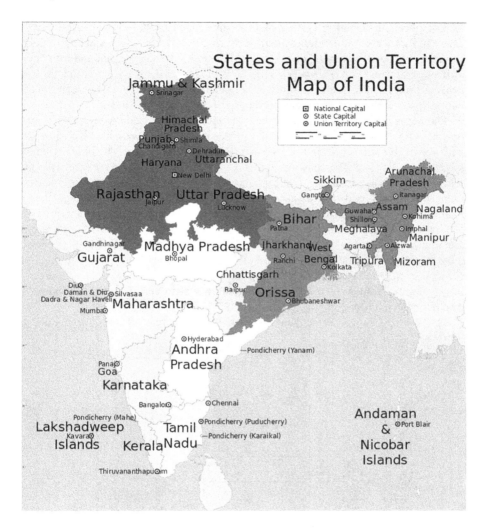

FIGURE 4.9 Indian map showing the black out (refer shorturl.at/dhB49).

4.10 VOLTAGE ISSUE IN POWER DISTRIBUTION NETWORK WITH RENEWABLE ENERGY INTEGRATION

The output power of the solar PV is extremely random and volatile, so when the quantity of solar PV in the distribution network is high, it will create power quality problems such as harmonics, flicker, voltage fluctuations, and voltage limits which further can be dangerous for power system safety and security. Due to this, the authorities limited the usage of distributed solar PV power in the distribution power network. The output of the solar PV is maximum in day time and zero in night time. In the modern power system, we have to focus on the problems of the distribution network while in traditional power system, we usually focused on the generation side.

The power distribution system requires integrated tools to perform efficiently. These tools are communication system, various control technology and coordination systems over the distributed generation, and the effective supervision and maximum utilization of generated renewable energy to decrease the difference between the load peak and load valley. It also improves the operational efficiency of the power distribution system to meet the modern power distribution system criteria.

The capacity of the electrical power supply can be improved through the integration of RES. But the instability in the power distribution network also increased with renewable energy resources. Generally, the capacity ratio of the distribution system is higher, but RES may affect this due to uncertainty of power insertion. The reliability, security, economy, and certainty of the power distribution system can be improved through real-time assessment. The fluctuations in the load voltage depend on uncertain factors and affect the voltage control. The integration of the distributed generation in the power distribution system significantly changes the voltage and current characteristics due to variable power injection, and it makes the voltage control more complex. The optimum operation of the integrated grid and DG try to meet the demand of the load power by regulating the power supply voltage. The characteristics of the bus voltage of the power distribution system are analyzed with variable distributed generation [28–30].

The voltage control strategy is represented in Figure 4.10 and is explained as follows:

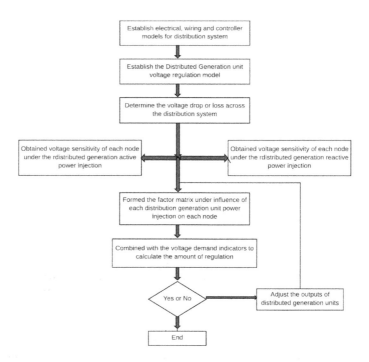

FIGURE 4.10 Process of voltage control strategy.

Voltage Issues in Power Networks

1. Use the simulation software to analyze the power distribution system of hybrid microgrid model, mainly for active power, reactive power, and voltage profile investigation.
2. The voltage control model of the power distribution system of distributed generator integration such as fixed power factor control, or constant voltage control.
3. The basic analysis of RES based on the voltage control after the active and reactive power injection is done for each node.
4. Create voltage sensitivity matrix for each junction and RES coupling point into the active power and reactive power.
5. Formed the objective function to optimize the control strategy of voltage requirement index.
6. Last, to recalculate the voltage of power distribution system, if it doesn't fulfill the requirements to voltage control, then modified the objective function or calculate the voltage again.

4.10.1 VOLTAGE ISSUES WITH SOLAR PV

The grid integration of solar PV allows the distributed solar PV networks to inject their generation into the national utility grid. It improves the economics of the solar PV by maximizing the usage of generated power and optimizes the energy balance by reducing the operational cost and providing the two-way connection between the user and the grid utility. The integration of solar PV and main grid is a very common connection throughout the world because the demand of the electrical energy is continuously increasing, but the reserve of fossil fuel is decreased, so we have to utilize the maximized renewable energy resources. The global installed capacity of the solar PV has seen exponential growth in previous years and reached to 631 GW in June 2020 and 36 GW in India in June 2020. According to the "renewable energy capacity statistics", China has the largest installed capacity of solar PV.

The advanced inverter technology, solar PV protection technology, grid protection technology, antiislanding technology, and solar PV forecasting technology are required to integrate the solar PV with the main grid. The range to the advance solar PV inverter is from light-duty inverter to the heavy-duty inverters for continuous output. The net metering is the difference between the power generated by solar PV and the load demand. If the generated power is in excess as compared to the load demand, the excess power is sent to the main grid and vice versa. Therefore, the antiislanding protection is required to immediately stop the flow of power when the voltage of the main grid goes down. As per the IEEE 1547 Section 4, "PV system power must be de-energized from the grid within two seconds of the formation of an island" [31]. So, it is necessary that PV plant detect the condition of island and stop the flow of power to the grid in minimum time. The maximum time is 2 seconds after the formation of island microgrid. The reconnection time of the inverter must be set at more than 60 seconds after the grid energized to avoid the unnecessary disturbance, and it can be done using the automatic island detection controls. A protective device is also used to monitor all the parameters relevant to the grid and disconnects the solar PV from the grid in case of any faulty scenario. This protective device is not required if the solar PV plant is working in islanding mode. Generally, the modern solar PV inverter has this type of protection system [30,32].

The management of renewable energy resources and forecasting technologies focuses on the measuring weather conditions as per NREL. The smart grid is an intelligent system which consists of multiple sensors and controllers to make a decision. It is capable to sense and manage the system overload, and also manage the congestion by rerouting the power to reduce the probability of power outage. The smart grid technology is more cost-effective as compared to traditional technology due to automatic controls and sensors if the distribution system has renewable energy integration. The DG's penetration in the power system can be of three different levels: low, medium, and high. The low penetration has a range between 1% and 5%, the medium penetration has a range between 5% and 30%, and more than 30% is known as high penetration of renewable energy. According to the 'Singapore Energy Market Authority report (2011)', Pulau Ubin, a small island was used to install the pilot project of microgrid which consists of solar PV system. This microgrid was very useful to perform the experiments and to develop the advanced technologies related to the microgrid for future projects. The main concept of the smart grid technology is the integration of distributed sources into the power distribution network. But the integration of renewable energy has its own challenges due to the limited and variable power output [33,34].

4.10.1.1 Technical Challenges, Economical Benefits, and Environmental Impact of Solar-Grid Integration

In the convention power system, the power flows in a single direction, i.e., from generators to the users through substations while the power can flow either side with the integration of solar PV. The problem is that the mostly power distribution networks were not adequate to either side power flow. The small mismatch between the solar PV generation and the load creates significant impact on the distribution feeders. When the solar PV generates more power than the load demand, the energy will start to flow toward the substation through the distribution feeder. It can induce reverse current and may disturb the grid parameters. The large-scale solar PV system requires separate transmission network to transmit their generated power which will increase the line losses.

The major challenges in installation of solar PV system are integration of solar and grid, frequency stability, and power quality. The large-scale distribution system has connected average load of more than 10 MW. The large-scale solar PV experienced the power quality issues.

It is not possible to share the land with agricultural farms in case of solar PV, like wind turbines. However, the solar plant can install at low quality areas to minimize the cost of the land. The solar PV can be installed in brown fields, mining lands, on river or lake, etc. The PV cell manufacturing is the complicated process which has involvement of a number of hazardous materials to clean the semiconductor surface such as hydrochloric acid, sulfuric acid, nitric acid, hydrogen fluoride, and acetone. The quantity of cleaning chemical is depending on the size of silicon wafer and type of cells. Workers also face risks associated with inhaling silicon dust. Therefore, the strict law is required to the safety of workers who are involved in manufacturing of solar panels and to dispose the waste products properly [34–36].

Voltage Issues in Power Networks

4.10.2 Voltage Issues with Wind DG

The single unit of wind turbine can be large up to 5 MW, so the issues of voltage quality are severe. The wind turbine has very high source impedance which is connected with the power distribution system. The supply current is distorted sometimes due to the presence of the harmonics which was introduced in the system due to power electronic converters. The power electronic converters, controllers, and drives in electric furnace, SMPS, and computer systems inject the harmonics and reactive power in the power distribution network. The distribution transformer requires reactive component of the current to fulfill the demand of magnetizing current. The requirement of electric power can't fulfill easily by increasing the production due to environmental issues, increasing cost, etc. It will decrease the power quality and impose the burden on the distribution network. This results in the reduction in the frequency and further decreases the system voltage [37].

The researchers are trying to improve the power quality. The authorities are using the active filters and compensators and also put the penalties to the users with low power factor or distorted voltage profile. The power system stability may improve using fast reactive compensators. The vector product of the supply voltage and the phase current defines the reactive power, for a single-phase AC system. The nonlinear loads have separate definition of the reactive power because their power factor is less than 1 but the voltage and current have no phase difference. The nonlinear loads have also separate definition of the power factor. The nonlinear load has two parts of overall power factor, the distortion pf and the displacement pf.

The displacement power factor is explained as "the cosine of phase shift angle between fundamental supply current and voltage".

$$\text{DF} = \sqrt{\left(\frac{\text{Sum of square of amplitudes of all harmonics}}{\text{square of amplitudes of fundamental}} \right)} \times 100\%$$

$$\text{DF}\left(\text{for rated current}\right) = \frac{\sqrt{\sum_{h=2}^{\infty} I_h^2}}{I_1} \tag{4.39}$$

4.10.2.1 Issue of Voltage Variation

The output of the wind turbine mainly depends on the direction and velocity of the wind, so the output power of the wind turbines may vary. It also causes the voltage fluctuations if the system has maximum dependency on wind turbines. Also, it generates the flicker effects during normal operating conditions. The cause of the voltage variations can be the variation in the load or the power generated by wind turbines when the wind turbine is integrated with the main grid at a particular fix speed. The low-capacity wind turbine faces more voltage fluctuations as compared to the large-size wind turbines during the short-time disturbances. The speed regulation range is wide and smooth for large wind turbines as compared to small which will provide a better regulation [38,39].

4.10.2.2 Voltage Dips Issue

The voltage dip is the reduction in the terminal voltage in the range of 1%–90% of the rated voltage in a very short time, generally 1 millisecond to 1 minute. According to the IEC 61400-3-7 standard, it is the "assessment of emission limit for fluctuating load". The system voltage reduces suddenly on the starting of wind turbines. The switching operation of wind turbine causes percentage change in the voltage which is given as

$$d = 100 K_u (\Psi_k) \frac{S_h}{S_k^*} \qquad (4.40)$$

where

d – relative change in the voltage,
$K_u(\Psi_k)$ – the factor to represent the change in voltage,
S_h – rated apparent power of the wind turbine and
S_k^* – short circuit apparent power of the main grid.

The maximum permissible limit of the voltage dip is $\pm 3\%$.

4.10.2.3 Harmonics

The nonlinear load is the main reason of the harmonic's distortion in the distribution network such as induction motor at low load, arc discharge lamps, magnetization of the core, electric arc furnaces, and distorted current. This current leads to the harmonics in the distribution network. The impact of the presence of the harmonics is harmful for the system. It will increase the power losses, low power quality, humming noise, and disturbance in the nearby communication system. The harmonic current and voltage must be in permissible limit at the moment of connection of wind turbines in the distribution network. The passive resonant filter consisting of LC has been used to reduce the harmonics in the system. The passive filter has its own limits, i.e., large size, resonance, etc. Therefore, the active filter can also be used, but the cost of active filter is high as compared to the passive filter. As per standard IEC 61400-21 guidelines, "harmonic measurements are not required for fixed speed wind turbines where the induction generator is directly connected to grid; harmonic measurements are required only for variable speed turbines equipped with electronic power converters".

Generally, power electronic controllers of the wind turbines are PWM inverters, which have carrier frequencies within the range of 2–3 kHz and generate the harmonic currents. The wave type grid voltage is not the sinusoidal voltage and not harmful for the life of electrical equipment. The odd integer harmonics of 5th, 7th, and 9th orders are usually present in the terminal voltage of the grid.

It is better to include the compensator while designing the power plant to avoid unnecessary distortions. It will also tend the voltage into permissible limits [40].

4.10.2.4 Flickers

Flicker may limit the peak power generation through wind turbine which is integrated to the main grid. The main cause of flicker is voltage fluctuations, which are induced due to sudden change in the load power. Wind turbine characteristics

Voltage Issues in Power Networks

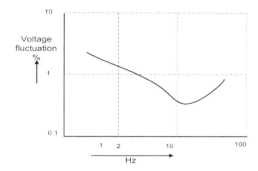

FIGURE 4.11 Effect of frequency variation on voltage fluctuations.

and grid conditions also affect the flicker emission of grid-connected wind turbines. Flicker will generate more disturbance in fixed speed wind turbine while less disturbance in variable speed wind turbines. The power output of the wind turbine will reduce as the rotor blade passes near the tower. It will cause periodic variations in a frequency of approximately 1 Hz.

The researchers proposed several solutions to minimize the flicker effect produced by grid integration of wind turbines. The flicker level of the voltage depends on the different parameters, e.g., amplitude, shape of turbine, and repetition frequency. The flicker meter responds to the frequency variation which is described in Figure 4.11.

4.10.2.5 Wind Turbine Location

The impact of integration of wind turbine into the distribution system depends on the method of connection. The power quality of the wind turbine is high when the WT is installed near to the load, and power quality is low when WT is installed far away. The existing customers feed through the wind turbine, which was already connected to the existing transmission line as shown in Figure 4.12. Generally, with the distance of few kilometers between the wind turbine and PCC, this connection is easy and cheap and affects the power quality on consumer's load [41].

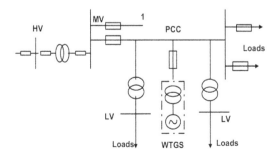

FIGURE 4.12 WGTS connection with MV.

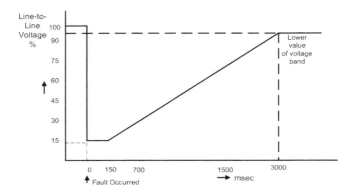

FIGURE 4.13 Low-voltage ride through capability.

4.10.2.6 Low-Voltage Ride through (LVRT) Capability

The wind turbine penetration has sufficient effect on the power system if the size of wind turbine is large. The maximum size of the wind turbine which does not affect the stability of the power system operation depends on the connected equipment's technology. The system must have protective equipment to reduce the harmful impacts of the voltage flicker, voltage variation, reactive power variation, frequency variation, etc. due to the wind speed variation. The contingencies of the power system support the critical issue of voltage collapse and shortage of reactive power. The LVRT capability is the most necessary equipment shown in Figure 4.13. The modern inverter has this LVRT capability but its cost is higher. At present, 38.9 GW power is generated through wind farms in India (February 2021). The generation through wind energy may be increased in upcoming years due to on-going projects [42–44].

4.11 CASE STUDIES/SIMULATIONS

In last few years, the installed capacity of the renewable energy increased significantly. 24.2% of power requirement of the Indian electricity sector is fulfilled by the RES at present. A number of solar power plants with bio-power plants were installed in many remote locations through the financial assistance of ministry of nonrenewable energy (MNRE) and rural electrical corporation (REC).

A number of central electrification schemes were started by the Indian government based on micro-grids, even for "remote" locations in the past few decades. However, most of them were failed because of on-ground challenges, despite of funding supports. This may have occurred due to lack of problem identification including societal requirements.

At the same time, in the Indian context, only 863 villages from 602,763 villages are not electrified, according to the GARV website. According to the current definition of electrification, "Having more than just a single wire in the village connected to the grid" [45]. But as per new government regulations, "Minimum 10% houses must be connected with the grid electrical supply for a village to be called electrified" [45]. However, villages are just one part of the problem; it is the users that are

Voltage Issues in Power Networks 101

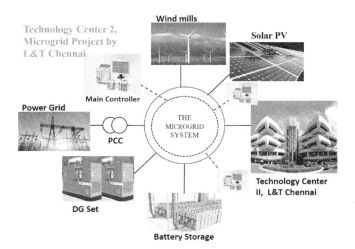

FIGURE 4.14 Hybrid microgrid installed in Chennai by L&T. (L&T construction.)

the real challenge for connectivity and quality supply. While quality should include issues like voltage and frequency, at the very least it should start with not being load shed. India is having the world's highest solar intensities with an annual solar energy yield of 3,500–3,800 kWh per kWp of the installed capacity; the Jawaharlal National Solar Mission has the aim of deploying 50 GW of grid-connected solar power by 2022 and aims at reducing the cost of solar PV power generation in India (Figure 4.14) [46].

The L&T developed the hybrid microgrid in its headquarter of construction's campus at Manapakkam in Chennai. The Engineering design and Research Center and other centers fulfill the conditions of green building standards and norms. The first solar was installed in 2008 and in operation since 2009. The entire campus of L&T has 1 MW rooftop grid-connected solar PV systems. The 8.7 MW wind farm was also designed and installed by L&T in Tamil Nadu to fulfill the demand of electricity in campus [47].

Technical characteristics of the installed microgrid

- 131 kWp solar PV
- 8.7 kW micro-wind power
- 10 kW/32 kWh Li-ion battery to store the energy and provide backup
- 2*808 kW diesel generators

In the MATLAB (Figure 4.15), the author models the hybrid microgrid with standard IEEE 13-bus distribution system in which two solar plants, one wind turbine, one battery bank, and multiple loads are modeled. The feeder has constant current load in the center. The model represents the power flow in the system 24 hours in a day. Based on the available technical data, the voltage of LV side of the transformer (110/33 kV, 12 MVA) was maintained at 32.6 kV with OLTC feature for constant voltage. The wind turbine and solar PV are interfaced with the power electronic

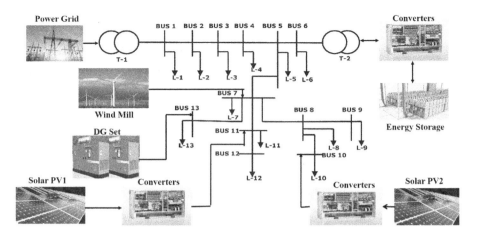

FIGURE 4.15 IEEE 13 bus distribution system for microgrid.

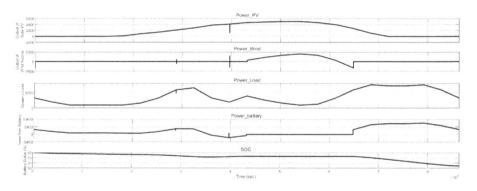

FIGURE 4.16 Voltage profile control of hybrid microgrid with distributed generation.

converters to provide the constant voltage and frequency, and the range of the power factor is 0.9–1 lagging or leading. The feeder has 28 km length, the minimum load is 287 kW, and the maximum load is 3128 kW.

The voltage profile of the simulated hybrid microgrid is shown in Figure 4.16. The voltage fluctuates with the insertion of the RES in the hybrid microgrid, but it will stable in the friction of milliseconds. When the power factor of the wind turbine is controlled, the production of the active power increased. The maximum voltage drop will occur at the feeder which doesn't include the generation unit. That case was modeled and studied to consider due to the possibility of a wind turbine stopped at full output power and maximum loading condition. If the voltage of the distribution system is reduced to 27.2 kV, the maximum permissible production is 3 MW. In that case, the voltage of the system feeder is at the permissible limits. If the wind turbine stops in that case, the voltage of the second feeder is also at the allowed limits which is the optimum condition for the system.

Voltage Issues in Power Networks

4.12 SUMMARY

The chapter provides the technical challenges regarding the voltage issues with the integration of RES into the power distribution network. Today, worldwide, the integration of high penetration level of RES into the existing power distribution system has a significant impact on the operation of the power system. Voltage collapse and the impact of renewable energy integration on various prospects were discussed in detail in this chapter. The various conditions and cases were discussed. The strategy to control the voltage using OLTC is discussed in detail. The effect of insertion of solar PV and wind turbine in the power system was discussed. A case study of hybrid microgrid installed in Chennai was discussed in detail. A model of 13-bus hybrid microgrid with solar PV, wind, and battery was simulated on MATLAB, and the results were obtained and discussed in this chapter.

REFERENCES

1. A.R. Bergen and V. Vittal, *Power System Analysis*, Prentice Hall, Hoboken, NJ, 2000.
2. A.Q. Al-Shetwi, M.A. Hannan, K.P. Jern, M. Mansur, and T.M.I. Mahlia, Grid-connected renewable energy sources: Review of the recent integration requirements and control methods. *Journal of Cleaner Production*, vol. 253, p. 119831, 2020. ISSN 0959-6526.
3. W.G. Chaminda Bandara, G.M.R.I. Godaliyadda, M.P.B. Ekanayake, and J.B. Ekanayake, Coordinated photovoltaic re-phasing: A novel method to maximize renewable energy integration in low voltage networks by mitigating network unbalances. *Applied Energy*, vol. 280, p. 116022, 2020. ISSN 0306-2619.
4. CIRED preliminary report of CIRED Working Group 04, Dispersed Generation Issued at the CIRED Conference in Nice, June 1999.
5. A. Schweer, Special report session 3. In *Proceedings of CIGRE Symposium on Impact of Demand Side Management, Integrated Resource Planning and Distributed Generation*, Neptun, Romania, 1997.
6. A.J. Petrella, Issues, impacts and strategies for distributed generation challenged power systems. In *Proceedings of CIGRE Symposium on Impact of Demand Side Management, Integrated Resource Planning and Distributed Generation*, Neptun, Romania, 1997.
7. A. Borbely and J.F. Kreider, *Distributed A New Paradigm for the New Millenium*, CRC Press, Boca Raton, FL, 2001.
8. IEA, *Renewable Energy, Market and Policy Trends in IEA Countries*, International Energy Agency, France, 2004.
9. S. Rahman, Green power: What is and where we can find it? *IEEE Power and Energy Magazine*, vol. 1, pp. 30–37, January-February 2003.
10. M. Thomson, Automatic voltage control relays and embedded generation. Part I. *Power Engineering Journal*, vol. 14, pp. 71–76, April 2000.
11. C.W. Taylor, *Power System Voltage Stability*, Mc Graw Hill, New York, 1994.
12. B. Milosevic and M. Begovic, Capacitor placement for conservative voltage reduction on distribution feeders. *IEEE Transactions on Power Delivery*, vol. 19, no. 3, July 2004.
13. V. Borozan, M.E. Baran, and D. Novosel, Integrated Volt/Var control in distribution systems. In *Proceedings of IEEE Power Engineering Society Winter Meeting*, 2001.
14. A. Augugliaro, L. Dusonchet, S. Favuzza, and E.R. Sanseverino, Voltage regulation and power losses minimization in automated distribution networks by an evolutionary multiobjective approach. *IEEE Transactions on Power Systems*, vol. 19, no.3, August 2004.

15. R.H. Liang and C.K. Cheng, Dispatch of main transformer ULTC and capacitors in a distribution system. *IEEE Transaction on Power Delivery*, vol. 16, No. 4, October 2001.
16. S. Corsi, P. Marannino, N. Losignore, G. Moreschini, and G. Piccini, Coordination between the reactive power scheduling function and the hierarchical voltage control of the EHV Enel system. *IEEE Transactions on Power Systems*, vol. 10, no. 2, May 1995.
17. H. Vu, P. Pruvot, C. Launay, and Y. Harmand, An improved voltage control on large-scale power system. *IEEE Transactions on Power Systems*, vol. 11, no. 3, August 1996.
18. B. Cova, N. Losignore, P. Marannino, and M. Montagna, Contingency constrained optimal reactive power flow procedures for voltage control in planning and operation. *IEEE Transactions on Power Systems*, vol. 10, no. 2, pp. 602–608, May 1995.
19. C.M. Affonso, L.C.P. da Silva, F.G.M Lima, and S. Soares, MW and MVar management on supply and demand side for meeting voltage stability margin criteria. *IEEE Transactions on Power Systems*, vol. 19, no. 3, pp. 1538–1545, August 2004.
20. P. Kundur, IEEE CIGRE joint task force on stability terms and definitions, definition and classification of power system stability. *IEEE Transactions on Power Systems*, vol. 19, no. 2, pp. 1387–1401, May 2004.
21. A.E. Hammad and M.Z. El Sadek, Prevention of transient voltage instabilities due to induction motor loads by static VAR compensators. *IEEE Transactions on Power Systems*, vol. 4, no. 3, pp. 1182–1190, August 1989.
22. M.H. Abdel-Rahman, F.M.H. Youssef, and A.A. Saber, New static var compensator control strategy and coordination with under-load tap changer. *IEEE Transactions on Power Delivery*, vol. 21, no. 3, pp. 1630–1635, July 2006.
23. A.C. De Souza, C.A. Canizares, and V.H. Quintana, New techniques to speedup voltage collapse computations using tangent vectors. *IEEE Transactions on Power Systems*, vol. 12, no. 3, pp. 1380–1387, 1997.
24. M. Moghavvemi and O. Faruque, Real-time contingency evaluation and ranking technique. in *IEE Proceedings Generation, Transmission and Distribution*, 1998. IET.
25. G. Verbič and F. Gubina, A novel concept for voltage collapse protection based on local phasors. In *Transmission and Distribution Conference and Exhibition2002*: *Asia Pacific*. IEEE. 2002. IEEE.
26. N.V. Acharya and P. Rao. A new voltage stability index based on the tangent vector of the power flow jacobian. In *2013 IEEE Innovative Smart Grid Technologies-Asia (ISGTAsia)*. 2013. IEEE.
27. M. Farrokhabadi, Automated topology processing for conventional, phasor-assisted and phasor- only state estimators. 2012, Master's thesis, Royal Institute of Technology (KTH).
28. C. Qian, Z. Wang, and J. Zhang. A new algorithm of topology analysis based on PMU Information. In *2010 5th International Conference on Critical Infrastructure (CRIS)*, 2010. IEEE.
29. A. Wood, B.F. Wollenberg, and G.B. Sheblé, *Power Generation, Operation and Control*. Wiley, Hoboken, NJ, 2013.
30. https://www.larsentoubro.com/ (L&T Construction for Case study)
31. https://www.nrel.gov/docs/fy16osti/67185.pdf.
32. M.M.M. Kamel, Development and Application of a New Voltage Stability Index for On-Line Monitoring and Shedding, M.Sc Thesis, The University of Tennessee at Chattanooga Tennessee, 2016.
33. F.A. Viawan, Voltage Control and Voltage Stability of Power Distribution Systems in the Presence of Distributed Generation, Ph.d Thesis, Chalmers University Of Technology Göteborg, Sweden, 2008.
34. E.G. Brown Jr., Governor, Microgrid Analysis and Case Studies Report, Energy Research and Development Division, California, August 2018.

35. C.H. Wei, A.I. Xin, W.U. Tao, and L. Hui, Influence of grid-connected photovoltaic system on power network. *Electric Power Automation Equipment*, vol. 33, no. 2, pp. 26–32, 2013.
36. R. Yan and T.K. Saha, Investigation of voltage stability for residential customers due to high photovoltaic penetrations. *IEEE Transactions on Power Systems*, vol. 27, no. 2, pp. 651–662, 2012.
37. R.A. Shayani and M.A.G. de Oliveira, Photovoltaic generation penetration limits in radial distribution systems. *IEEE Transactions on Power Systems*, vol. 26, no. 3, pp. 1625–1631, 2011.
38. Y. Wang, F. Wen, B. Zhao, and X. Zhang, Analysis and countermeasures of voltage violation problems caused by high-density distributed photovoltaics. *Proceedings of the CSEE*, vol. 36, pp. 1200–1206, 2016.
39. Y. Che, W. Li, X. Li, J. Zhou, S. Li, and X. Xi, An improved coordinated control strategy for PV system integration with VSC-MVDC technology. *Energies*, vol. 10, no. 10, p. 1670, 2017.
40. N. Jayasekara, P. Wolfs, and M.A.S. Masoum, An optimal management strategy for distributed storages in distribution networks with high penetrations of PV. *Electric Power Systems Research*, vol. 116, pp. 147–157, 2014.
41. W. Zhang, Z. Liu, and L. Shen, Flexible grid-connection of photovoltaic power generation system with energy storage system for fluctuation smoothing. *Electric Power Automation Equipment*, vol. 33, no. 5, article 106G111, 2013.
42. Q. Li and J. Zhang, Solutions of voltage beyond limits in distribution network with distributed photovoltaic generators. *Automation of Electric Power Systems*, vol. 39, no. 22, pp. 117–123, 2015.
43. K. Ma, C. Yuan, J. Yang, Z. Liu, and X. Guan, Switched control strategies of aggregated commercial HVAC systems for demand response in smart grids. *Energies*, vol. 10, no. 7, p. 953, 2017.
44. X. Xu, Y. Huang, C. Liu, W. Wang, and Y.L. Wang, Influence of distributed photovoltaic generation on voltage in distribution network and solution of voltage beyond limits. *Power System Technology*, vol. 10, no. 34, pp. 140–146, 2010.
45. http://www.ddugjy.gov.in/page/definition_electrified_village.
46. S.W. Mohod and M.V. Aware, Power quality and grid code issues in wind energy conversion system, in *An Update on Power Quality*, D.D.-C. Lu (ed.), IntechOpen, 2013. doi: 10.5772/54704.
47. K.N. Nwaigwe, P. Mutabilwa, and E. Dintwa, An overview of solar power (PV Systems) integration into electricity grids. *Materials Science for Energy Technologies*, 2019. doi: 10.1016/ j.mset.2019.07.002.

5 Reactive Power Management in Power Systems Integrated with Renewable Generations

Satish Kumar
KIET Group of Institutions

Ashwani Kumar
NIT Kurukshetra

CONTENTS

5.1 Introduction: Importance of the Presence of Reactive Power 108
5.2 Purpose of Reactive Power Management 108
5.3 Management of Reactive Power with Conventional Methods 110
5.4 Reactive Power Assessment of Integrated Power System for RES 110
5.5 Reactive Power Planning and Assessment Schemes in Present
Electricity Market 111
5.6 Current Trend of Renewable Energy Resources (RES)............................... 112
    5.6.1 Contribution of Renewable Energy Resources (RES) in
Modern Electricity Market 112
    5.6.2 Reactive Power Issues and Assessment with RES 113
    5.6.3 Modern Intelligent Techniques/Tools for Reactive Power
Management 114
    5.6.4 Contribution of Reactive Power Management in the
Operation of Modern Power Systems 116
    5.6.5 Novel Methods and Latest Trends in Reactive Power
Control and Management 117
        5.6.5.1 Requirement of Reactive Power Management................... 117
5.7 Sensitivity Analysis-Based Reactive Power Control 118
5.8 Modeling and Implementation of FACTS for Reactive Power Management 119
5.9 Modeling of FACTS Devices for WIS with Firing Angle Control.............. 120
5.10 Cost Analysis-Based Reactive Power Compensation Using Intelligent
Techniques 120

DOI: 10.1201/9781003271857-5

5.10.1 Salient Pole Rotor Synchronous Generator 123
5.11 Synchronous Machine as a Synchronous Condenser 124
5.12 Conclusion 126
References 127

5.1 INTRODUCTION: IMPORTANCE OF THE PRESENCE OF REACTIVE POWER

The reliability facilitates active commercial transactions across all transmission line networks and can be easily supported with the help of voltage control and reactive power management technique having single-line activity. By managing production and reactive power absorption, voltage in the AC system can be easily controlled. The importance of management of reactive power can be summarized by three different cases.

In the first case, both power system and customer equipment are modeled to work within the fixed and wide range of voltages say, $\pm5\%$–10% of the normal range of voltage. We observe that the performance of a large number of equipment like induction motors, light bulb, and many more deteriorates at low levels of voltage. Similarly, high voltage causes damage in the equipment/its part, thereby reducing its efficiency and life.

The main source of reactive power consumption comes from the system of generation and transmission. For maximizing real power which can be transferred in a congested transmission system, reactive power flow must be controlled as well as optimized.

The main source of real power is the movement of reactive power on the transmission loss. Hence, provision of additional implementation of energy and capacity must be there to compensate for the incurred losses.

5.2 PURPOSE OF REACTIVE POWER MANAGEMENT

To maintain the voltage throughout the system, SVC, synchronous generators, and various types of distributed energy resources are used to increase the voltage injection of reactive power into the system, whereas absorption is done to lower the voltage levels. Research shows that voltage support in the system is a function of three parameters, i.e., magnitude and location of generator output with the load on customer side along with the configuration of the distributed energy resources.

Three important objectives of managing reactive power can be summarized as follows:

- It should be capable of providing sufficient voltages through the distribution and transmission system to meet contingency and current conditions.
- For real power flows, congestion minimization must be done.
- Real power loss must be minimized.

In general, reactive power control is performed using the following:

- Control of excitation
- Reactors and switching of shunt capacitor banks

Reactive Power Management in Power Systems

- FACTS devices and VAR methods
- Transformers of tap-changing and regulating types.

The important and main part of the system planning is identification of problem of optimal volt-ampere reactive (VAR) sources in the transmission and distribution systems for its efficient operation. The sole objective of planning of VAR is to compensate for the reactive power requirement for system expansion and improvement in the voltage profile of the system considered [1] and to reduce real losses within the system. As a case study, the objective function may be formed comprising of cost of installed source of reactive power and reduced real loss of the system or the combination of both the factors. By taking into consideration the possible contingency condition and projected load expansion in future, the voltage profile of various buses may be improved for stabilizing voltages. Further optimal distribution and allocation of reactive power, proper maintenance, and improvement in the voltage profile may be implemented in the transmission line. Synchronous generators are the main and feasible source of generation of reactive power into the transmission system [2]. The system consists of various loading conditions and reactive power requirement of these loading conditions, and some additional sources like capacitor bank may also be used as they are capable of injecting reactive power to maintain and improve the voltage profile of these loading conditions. The problem of planning of optimized VAR sources may consist of minimization of typical system losses in the system considered with few loading conditions with some investment constraints [3], balance equations of power flow solution, and constraints of security for all the loads. Reactive power control and voltage control are important for the proper operation based on day-to-day load requirements at the substation level in the transmission system. Voltage deviations, with system losses, may be reduced by optimized management of reactive power. Therefore, voltage and reactive power must be controlled effectively and efficiently to manage the balanced voltage profile with lower power generation cost involved. It can be summarized that generation of reactive power from two sources, i.e., condensers and synchronous generators, is very effective [4] for controlling the voltage profile in the transmission system.

In vertical electrical power system model, reactive power management was provided by the dispatch centers, and there was no proper mechanism to charge the reactive power except the KVA demand charge taken from the industries. However, in the current competitive electricity market, reactive power deployment is carried out by the system operator, and this system operator arranges the reactive power from either condensers or the shunt FACTS devices [5,6]. Considering the higher reactive power demand within the system depending on the reactive load requirements, the generators may require generating more reactive power at the cost of active power reduction which causes the loss opportunity cost [7]. Since the huge cost is involved for the reactive power generation from the condensers or synchronous generators or FACTS devices, the reactive power providers need to be remunerated based on the supply of reactive power. Thus, there is a need for the transparent pricing structure of reactive power along with the real power price. There must be some mechanism to obtain the reactive power cost [8,9].

5.3 MANAGEMENT OF REACTIVE POWER WITH CONVENTIONAL METHODS

The importance of reactive power comes into play when addressing the grid instability w.r.t the voltage, also known as voltage instability. Hence, the effective compensation and implementation is required to restrict the circulation between the source and the load. This compensation is also required for power factor regulation as well as to ensure voltage stability of the system considered. The regulation of reactive power addresses the following factors of grid management [10,11]:

- Voltage instability
- Improved utilization of active power
- Power factor
- Power quality
- Improved system efficiency
- Adherence to the grid code [12].

Modern advanced Flexible AC Transmission System (FACTS) devices and technologies may also be used for effective reactive power factor compensation. In general, two types of compensation are employed, i.e., dynamic series and dynamic shunt. Being advantageous, the dynamic shunt-type compensation can automatically support the voltage profile in the particular area of the system considered. So, there is every reason to reduce unwanted reactive power [13]. There are three basic solutions for reducing reactive power [14]:

- Capacitor bank
- Static VAR generator
- Active dynamic filter/harmonic filter.

5.4 REACTIVE POWER ASSESSMENT OF INTEGRATED POWER SYSTEM FOR RES

Considering a large electrical interconnected power system, the point of collapse or the margin of voltage stability is preferred to determine some critical points in the system. As the system conditions get changed due to increase in load and other parameters like excitation/real/reactive power, the system condition also changes.

From the literature [15], it is seen that as the parameter of the system changes, the indication or the approximation of voltage stability cannot be assessed using the critical point method. In this case, the sensitivity analysis method is the best suited method to determine the point of disturbance, the critical point in the system, and the voltage collapse point in the system. Further, the voltage indices method or the method of sensitivity index are also very useful and effective to assess the closeness of point of unstable region. Several indices methods to solve the problem of instability are given in the recently published report by IEEE [16].

For the calculation of sensitivity for interconnected system, system stability index is identified for power flow analysis. Parameters like active power flow, reactive

Reactive Power Management in Power Systems

power flow, bus angle profile, and bus voltage profile are observed and determined for better assessment and planning of the system to avoid the chances of blackout or possible voltage collapse. Determination of bus sensitivity is important to avoid critical points in the system. The actual value of bus sensitivity in terms of voltage and angle magnitude helps in determining the actual point of possible voltage collapse or the point of voltage instability [17].

5.5 REACTIVE POWER PLANNING AND ASSESSMENT SCHEMES IN PRESENT ELECTRICITY MARKET

Proper monitoring and control of reactive power for smooth and lower power loss are important as well as challenging. The supply of reactive power must be installed to close vicinity of its consumption. Proper coordination and installation of dynamic and static voltage support are required to maintain the acceptable voltage limit and to avoid voltage collapse/voltage instabilities. The following are the key schemes for better and controlled reactive power assessment and planning in the electricity market:

- Voltage coordination
- Voltage schedule
- Transformers tap setting
- Reactive device settings
- Load shedding schemes
- Voltage and reactive power control schemes
- Reactive reserves for transient operating conditions
- Statics and dynamic voltage support
- Role of reactive power in ancillary services.

Ancillary services mean the different types of operation performed in the modern power system apart from transmission and distribution, and are important and mandatory to assess the grid security and stability of the power system. Ancillary services not only help maintaining the proper power flow within the system but also the proper coordination between supply and demand. After the power system failure, they improve or recover the health of the system. For the system where the integration of renewable energy is preferred with the conventional grid, they help to increase uncertainty and variability of the entire system. Synchronized regulation, contingency reserve, black start reserve, and flexibility reserve can be included in the ancillary services.

The recent report of Federal Energy Regulatory Commission (FERC) [18] has identified six ancillary services as given below:

1. Ancillary service of voltage and reactive power control
2. Ancillary service of compensation of losses
3. Ancillary service of dispatch as well as scheduling
4. Ancillary service of load flow
5. Ancillary service of protection of system
6. Ancillary service of maintaining energy balance.

Some common examples of these services are

- Frequency control
- Spinning reserve
- Operating reserve.

In Frequency Control Ancillary Services (FCAS), the operator at the grid side ensures short-term supply and asks for balancing of the supply for the rest of the electrical power system. This can be effectively controlled through the precise control of frequency via operational reserve which responds to system disturbances.

To maintain voltage stability limits within the prescribed limits, independent system operator is supposed to procure the special kind of ancillary service, i.e., reactive power. The general practice to procure these services is either from independent or affiliate generators using the voltage schedule method. Comparison with other ancillary services like regulation or reserve, where procurement is done based on market scenario, reactive power may not be available easily, rather it requires a specific process. The methods for procurement of reactive power are limited and nonexistent. The services of reactive power are not incorporated during the dispatch process but by the local and specific nature of the reactive power. Hence, it would be uneconomical to procure reactive power on a half-hour clearing market.

Ancillary service procurement in deregulated electricity market is very complex and tedious issue for independent system operator. The benefits in terms of service to the system include security, reliability, and economics. For better planning and assessment, various researchers have given methods for better procurement of reactive power support for ancillary services. Hence, procurement of reactive power support is made with the providers who give maximum societal advantages like high-marginal benefits with flexible price range.

5.6 CURRENT TREND OF RENEWABLE ENERGY RESOURCES (RES)

5.6.1 CONTRIBUTION OF RENEWABLE ENERGY RESOURCES (RES) IN MODERN ELECTRICITY MARKET

In the present electricity market, the impact of the various renewable energy resources like wind system is increasing rapidly and exponentially. Nowadays, conventional grid offers a variety of challenges such as increase in pollution, change in the climate conditions globally, and last but not least, the security issues associated with the supply of energy. At present, the growth rate of RES is about 20%, and this growth rate is increasing day by day as compared to any other energy resource. It is assumed that with this continuing growth rate, the wind energy system will be capable to generate about 12% of words total electricity by the year 2020 [19].

As the fastest and very common growing source of energy, wind energy system is becoming popular for generation of electricity. As the electricity generation by the wind energy systems also depends upon the fluctuating nature of the wind speed and the environmental conditions, it differs from the conventional method of generation. In the recent years, various types of wind energy system have been developed,

Reactive Power Management in Power Systems

installed, and integrated with the conventional grid to produce more secure and cost-effective electricity for the customers. At the same time, if we include the parameters like geographical location of wind systems with some electrical parameters like battery charging and mechanical parameters like water pumping, the generation with the wind system can be more profitable [20,21].

It is also seen that with the inclusion of some new control strategies in the interconnected multimachine system, the cost effectiveness in the generation of electricity with the wind system can be further increased. Another method may be inclusion of some dynamic controllers in the system to increase dynamic performance as well as ensuring dynamic stability of the system also. In the recent years, wind turbines with variable speed configuration of DFIG type are also becoming popular for advanced reactive power and voltage control for WIS. The integration of wind power with the conventional grid also causes some serious issues like stability and power quality. Higher level of wind penetration with fluctuating wind nature also affects the voltage profile and available power transfer capacity of the system.

Nowadays, newly designed wind energy systems which consist of DFIGs and located in between the distribution system, are becoming very popular. Recently, the wind turbines that consist of squirrel cage induction generator and DFIG are tested on IEEE 14-bus test system for small-scale signal stability limit. The comparison of both types of generation system has been done and studied. From the experimental data, it is seen that the operation of fixed-speed-type wind generators is easier, simpler, and reliable, but in terms of energy output by the wind system, they are limited. Apart from this, the results show that if the rating is the same, the variable-speed-type wind generators are more capable of improving stability of the system.

5.6.2 Reactive Power Issues and Assessment with RES

While addressing the issues of reactive power for wind-integrated system, it is observed that reactive power regulation capability cannot be provided by the simple induction-type generators [22]. When the system is operated continuously, a huge amount of reactive power causes deterioration and decrement in the voltage profile of the system due to system contingencies.

Due to the inherent voltage control capacity, DFIG-type wind turbines are better as compared to the wind turbines of induction generator type, for transient voltage stability characteristics. Further, for the same rating and capacity, the voltage recovery performance of the DFIG-type wind turbine is superior to induction generator type.

To increase the reactive power capacity of the wind-integrated system, some parameters of the induction generators like stator reactance, stator resistance, and rotor reactance have been proven very effective [23]. Wind turbine reactive power also depends on some exclusive parameters like resonant frequency of hub generator, coefficient of inertia, with some operation grid parameters like compensation of reactive power, location of fault, wind speed, duration of fault, and short-circuit power at the connection bus.

By calculating reactive and real power within the system, actual point of voltage collapse or critical point may be calculated. But for a large interconnected system, the actual value of this point may differ due to the actual operational values of the system. Hence, whenever a huge and very large system is taken for operation, sensitivity-based analysis is performed using stability index. Various stability parameters like point of collapse, actual critical point, reactive power, and many other operational variables are calculated using the stability index method.

While performing optimal placement of wind turbines in an integrated wind system, evaluation of some critical information such as frequency of wind direction, and wind speed should be monitored closely. So, when the integrated system is in operation, the information of the wind energy system should be in accordance with the actual operating condition of the system. The difference in the operating conditions and the actual conditions may lead to difference in the power generation by the system.

The purpose of the load flow studies is to ensure smooth and proper electrical power flow from generation to the customers by maintaining economical, reliable, and stable grid operation. Information and assessment like phase angle and magnitude at each bus with real and reactive power flow in the system may also be obtained with novel load flow techniques.

Out of the various techniques for load flow, two iterative, conventional, and popular methods, i.e., Gauss-Seidel (G-S) and the Newton-Raphson (N-R) techniques are available for the solution of the load flow problem. Out of these two, the NR method is better as compared to the GS method, due to owing of its faster convergence characteristic.

But the possibility of flat start is the only drawback of the conventional NR method as in the beginning the solution may oscillate without converging toward the solution. Hence, after performing few iterations, the load flow solution must be started with GS method and consequently NR method. The extension of NR method is the fast-decoupled method, which is more accurate, approximate, and faster as compared to the NR.

Just to meet the ongoing challenge of day-to-day higher power demand and security in the operation, the importance of assessment of reactive power for wind-integrated system has increased a lot. Out of the abovementioned techniques for load flow, various techniques of power flow like conventional power flow, continuous power flow, optimal power flow, NR power flow, and GS power flow are in vogue. The latest optimization techniques such as GA, PSA, APSA, and ANN are also useful to calculate voltage stability index, for enhancing margin of voltage stability and security of the interconnected power system for existing line loading conditions.

5.6.3 Modern Intelligent Techniques/Tools for Reactive Power Management

With the invent and development of some latest new software like power system analysis tool box (PSAT), OPELRT and GAMS, the problem of voltage stability can be addressed better by Power Flow (PF), Continuous Power Flow (CPF) and Optimal Power Flow (OPF) as compared by the conventional intelligence and computational techniques. The GUI feature of PF, CPF and OPF proved to be very useful in determining the margin of stability and point of collapse accurately. For a particular given load condition, the steady-state operation in normal working condition

Reactive Power Management in Power Systems

can also be determined and calculated with the help of, currents, voltages, reactive power and real power of the system considered. Hence, the purpose of power flow or load flow can be fulfilled for reactive power flow solution [24]. The method of CPF, an auxiliary method of power flow can also be used and well planned for better accuracy in the system for the assessment and locating of critical point in the system.

Out of the various power flow techniques, OPF method is best suited and accurate method. For an integrated system, the total cost of generation, is best determined by using OPF with the help of power balance and power flow equations at each node considering some of the inequality constraints like, operating limit of the system with control variables [25,26].

In the past, OPF problem only consists of systems like thermal energy and conventional grids with system parameters. In the recent years, with the development of a variety of new renewable energy resources, the classical problem of OPF includes the generation cost and recovery cost for using multiple FACTS devices. Hence, determination and evaluation of OPF with renewable energy systems and FACTS devices become challenging for integrated system. This integrated system with OPF solution is used to meet the increased power demand of consumer with increased and variable load. But stability of this interconnected system (voltage and thermal limits) should also be monitored for smooth power flow in the system.

It is seen that this challenge of maintaining stability limits is very well addressed and solved by OPF with proximities of voltage instabilities method [27]. That's why it is necessary to solve the power flow problems for base load as well as full load to keep the system within the stability limits. When considering the wind-integrated system, the solution may also include the variable parameter of wind system, i.e., the variable wind speed to execute economic performance of all generating units within the system.

PSAT is the additional toolbox for solving the problem of power flow, continuous power flow, and optimal power flow for complex and multimachine system. PSAT is free software developed by Fredric Milano and works with most of the MATLAB versions. The additional feature of PSAT is its GUI interface, which enables the user to plot for small signal stability analysis and various curves like P-curve, Q-curve, and many more to determine the stability and margin of stability of the system [17,28]. Network design using Simulink-based library of the PSAT and MATLAB is also very helpful. The power flow routine, which is the base of PSAT core, is used for initialization of state variable for complex problem design and solution. Based on the result of PF, CPF and OPF, static and dynamic characteristics of the system under study can also be obtained. Apart from mathematical routines and models, PSAT also offers a large number of user-friendly utilities [29] like

- Availability of Simulink library
- Availability of GUIs
- Installation and model construction by user
- Plotting results using GUI
- Easy conversion of data from one format to another
- Availability of command logs.

5.6.4 Contribution of Reactive Power Management in the Operation of Modern Power Systems

Generation: An increase in the generator output terminal voltage requires increase in the magnetic field (due to current in the field windings), which causes the development of reactive power. Thus, for limiting absorption of reactive power pattern, the magnetic field pattern of the stator is to be controlled. This prevents excessive heating at the stator and core ends. When absorbing a large amount of reactive power, synchronizing torque can also be reduced, which can further minimize the possibility of loss of synchronism.

Synchronous condenser: They are exclusive version of synchronous machines, which are specifically designed to provide reactive power in the structured network. They provide mechanical power to the load and hence can also be used as dynamic voltage support to the system. Synchronous condensers are capable of consuming a handsome amount of real power machine which is almost equal to 3% of machine reactive power.

Inductor and capacitors: Passive devices such as reactors (inductors and capacitors) are also responsible for generating and absorbing reactive power without any significant real power loss/additional expenses. For example, in case the voltage level goes down to 0.95pu, inductor and capacitor bank of around 100 MVAR rating may absorb (produce) 90 MVAR. At the same time, it can absorb (produce) 110 MVAR for a voltage rise of 1.05pu.

Static VAR compensators (SVCs): Combining inductors and capacitors, SVC possesses the ability of fast switching of the order of 1/60 second, thereby providing a continuous range of voltage control. This range can be extended for absorption and delivery of reactive power. Consequently, very fast and effective reactive voltage support can be obtained by redesigning the switching limit. SVCs mainly use capacitors, which causes deterioration in the reactive power due to lowering of voltages. They also require some unique harmonic filters to minimize the number of harmonics supplied to the system.

Static synchronous compensators (STATCOMs): The Family of FCATS devices consists of various power devices. STATCOM is a shunt device which is used for generation and absorption of reactive power. If comparison is made based on power electronics, response speed, and control capabilities, STATCOM is similar to SVCs. Similar to SVC, STATCOM is also used to provide very fast and effective voltage control. As far as short-term overload capacity of STATCOM is concerned, it does not suffer as seriously as SVCs. STATCOM is a current-limited device, and hence its MVAR capability responses linearly for voltages as compared to SVC, whose response is squared of the voltages.

Distributed generators: Distributed generators are very beneficial, if the generation can supply reactive power, failing which the performance of transmission and distribution is deteriorated. Induction generators are inexpensive in nature and hence are obvious choice for small grid-connected generation. They are also useful for renewable energy resources for reactive power generation and absorption. Distributed generators are very popular and important for the reactive power control. They are connected to the conventional grid with new and advanced solid-state

Reactive Power Management in Power Systems   117

power devices, so that the speed of the prime over may be synchronized with the system frequency. Nowadays, new hybrid energy storage devices like flywheels, superconducting energy storage devices and modern batteries are also connected to the conventional and smart grid to regulate reactive power control as in the case of STATECOM.

Transmission side: The prime source of reactive power in the system of transmission line is "unavoidable sequence of loads" along with phase shifting between current and voltages. This reactive power is used for controlling voltages of line. Sometimes the transmission line itself acts as a source of reactive power. It is also seen that a line without load, known as open line, acts as a capacitor and also acts like a source of leading capacitive or reactive power. Thus, the line becomes the source of lagging or inductive reactive power if it is conducting a high current.

5.6.5 Novel Methods and Latest Trends in Reactive Power Control and Management

For modern electrical power system stability improvement, it is always desired to lower the effect of reactive power within the interconnected system. Two types of electrical power are available with the use of AC system. First is real power, and the second is its byproduct reactive power. All the useful work in the system is governed by the use of real power, while the voltage is supported by reactive power in the system, and this voltage level must be controlled for system reliability as well as security. As the reactive power affects the entire voltage levels throughout the system, it has key significant and considerable effects on the system security [30].

5.6.5.1 Requirement of Reactive Power Management

For smooth and reliable operation of various equipment in power systems like motors and generators, it is important, but challenging, to prevent these equipment from damage and overheating, to avoid voltage collapse, and to ensure the stability and security limits of the system concern by proper monitoring and control of voltage.

Requirement of reactive power is validated if the drop in voltage is very low. In this situation, some generators connected in the system isolate themselves automatically to prevent voltage instabilities. Whenever there is abrupt increase in the load or when the generation is not up to the desired level, the drop in voltage occurs, which leads to the reduced reactive power from capacitors and line charging. Now if this reduction in voltage/voltage levels continues to reduce further, it may cause tripping of some additional equipment and loss in the load. Due to this, the entire system fails to provide reactive power demand within the system [31].

Synchronous condenser: Owing to the advantages of controllable power factor, system reliability, increased operating efficiency, and low sensitivity for variation in the voltage fluctuations, synchronous condensers are primarily used in complex and large power systems. Synchronous condensers find a variety of applications like power plants, compressors, mills, fans, and refineries because of their constant and controlled speed characteristics for reactive power compensation. The name "synchronous condensers" is because of their unique and variable external control mechanism when connected to the load. When the field excitation of synchronous condensers is in "under-excited" mode, they behave as a "variable inductor", and as "variable capacitor" when the field is operated on "overexcited" mode. In variable capacitor mode of operation, the synchronous condensers operate at leading power factor and at lagging power factor when operated as variable inductor. In general, for industrial applications where a large number of induction motors are employed, synchronous condensers are used and operated at leading power factor for power factor correction and compensation.

Capacitor bank: In modern electrical power distribution system, the main function of capacitor bank or static capacitors is to provide power factor improvement/correction. These capacitor banks are used to provide inductive loading from devices such as induction motors and other electrical drives operating on lagging power factor, and transmission lines, making the load to be resistive at its most. Another function of these devices is to control the voltage level at the customer end by controlling the voltage drop or reactive component generated by inductive reactive loads. They not only improve the efficiency of electric power system but also help in improving the overall stability of transmission line and distributed system during heavy load and disturbances.

The operation of capacitor bank during online monitoring and control is very stable and efficient. They can be connected online continuously to meet the requirement of steady-state reactive power of the distributed system. Additionally, they can be on/off to match dynamic reactive power requirement. Two types of reactive power compensation are applied in modern distributed power system: shunt type and series type. In a distribution system, shunt-type compensation is installed near the load along with distributed feeder. For better efficiency of the system, the capacitor bank is installed closer to the actual inductive load.

5.7 SENSITIVITY ANALYSIS-BASED REACTIVE POWER CONTROL

In reactive power control, the analysis of sensitivity for interconnected multimachine system is done with the use of present operating condition of the system considering all constraints. The following novel methods can be employed for sensitivity analysis:

1. Method of state indices [32]
2. Large deviation-based indices method
3. Method of voltage stability using generator sensitivity
4. Method of eigen value calculation
5. Method of tangent vector
6. Method of simulated RES model
7. Method of branch current calculation
8. Method of bus voltage magnitude and angle control [33].

Further, the following may also be used in association with the abovementioned submethods:

- Calculation of power flows repetition
- Method of continuation
- Method of optimization of variables

- Method of direct calculation
- Method of the closest distance to maximum transfer boundary
- Method-based energy function [34].

5.8 MODELING AND IMPLEMENTATION OF FACTS FOR REACTIVE POWER MANAGEMENT

Variable nature of wind is extremely challenging especially when the integration of the renewables is done with the convention electricity generation system to increase the economics of the generation [35,36]. The problem of variable wind speed can be addressed by the new and advanced power flow techniques like optimal power flow. The prime objective of optimal power flow is to curtail the total cost of generation using some power equation at every node in the system with some inequality constraints like operating limits of the network in terms of line flows and voltages and control variables [37,38]. Hence, optimal power flow is solved and executed to meet the current high demand of electricity and variable load by the customer side. One advantage of performing optimal power flow is to increase the voltage stability and thermal limits of the system [39]. Another advantage of performing optimal power flow is the intermittency in the power flow pattern due to voltage variation in the integrated transmission systems. Nowadays, in place of optimal power flow, a classical power flow consisting of generation cost and the cost of power devices [40] is designed and solved to solve day-to-day challenges in the generation and distribution system.

So one can conclude that though the maximization of generation is rigorous and tedious work, it can be held by tuning some of the power devices in the interconnected system using the latest optimization techniques and software like GA (Genetic Algorithm), PSO (Particle Swarm Optimization), APSO (Advance Particle Swarm Optimization), GAMES, and many more. Modern research shows that solution of optimal power flow problem using evolutionary programming is easier and user-friendly as far as the complexity and components of the system used. Apart from conventional problem, these optimization methods can also be used for optimization of the cost function of the system. Due to the absence of evolutionary operators like mutation and crossover [41] in GA, optimal solution by PSO is easier and advantageous as compared to GA. Application of PSO possesses the special feature of updation of particle (variables) itself due to inherent internal memory in it. Sometimes, to achieve direct power flow results in terms of voltage magnitude and angle, the calculation of power flow and optimal power flow is performed with the help of MATLAB toolbox, i.e., MATPOWER.

In a modern highly demandable environment, where the status of load is fluctuating and increasing, the applications of PSO can be further extended for integration and locating multiple FACTS devices into the system. The key purpose of installation of FACTS devices is to increase system loadability, improved voltage profile, and mitigation of various unwanted losses in the system. In the literature [42], it is evident that optimized installation with optimized placement of FACTS is best suited by PSO as compared to GA for increased availability of power demand at customer end. In the simulated model and advanced algorithms for controlling reactive power and modern control strategies, variable type speed turbines equipped with DFIG may also be used.

5.9 MODELING OF FACTS DEVICES FOR WIS WITH FIRING ANGLE CONTROL

With the invent of advanced power devices possessing a higher degree of regulating freedom for controlling various voltage levels and power flow in the integrated network, FACTS devices proved to be the key solution. They are also capable of providing controlled and noncontrolled power system mode of operation at certain optimized location of integrated power system as well as to increase available power transfer capacity with optimized generation and utilization. Modeling of FACTS controllers is also done and employed to incorporate the problems and optimal solution for the integrated interconnected systems.

Power flow solution with FACTS: Simultaneous and sequential solution methods are generally applied to get the solution of the power flow equation with multiple FACTS devices. The first method includes the calculation of some state variables of the interconnected system, i.e., bus voltage magnitude and angle with the use of the NR method. But in this method, the automatic updation of state variable is not possible, so it gives rise to the formation of new and special subroutine for updation of these variables.

The simultaneous method consists of solution of these state variables with controlled FACTS devices available in the interconnected system network. This method gives advantages of less convergence time as compared to the NR method of power flow solution and can be used for modeling of FACTS devices with solution of power flow problems.

5.10 COST ANALYSIS-BASED REACTIVE POWER COMPENSATION USING INTELLIGENT TECHNIQUES

Synchronous generator: When considering the problem of stability in the interconnected power system, i.e., voltage stability with long-term stability, the reactive power limit of the synchronous machine must be considered. The advantage of synchronous generators is that their output is rated as the maximum MVA output for fixed power factor and specified voltage. The range of the power factor may be from 0.85 to 0.95. With this limit of power factor, they can generate the output without violating the heating limits. The limit of active power output by synchronous generator is maintained by prime over to meet its MVA rating. The limit of uninterrupted power output is maintained by assuming three key factors [23,24]:

1. Limit of armature current
2. Limit of field current
3. Limit of end region heating.

For reliable and secure operation of the integrated interconnected power system, the provision and supply of reactive power is very important. In deregulated electricity market, the important task of the system operator is to provide reactive power support, and the cost of this support may be recovered by calculating price, based on the approximation method. Two different cases are possible using this method of costing. In the first case, the price of the active power includes the reactive cost and in the second one, calculation and inclusion of penalty factor with power factor is done to compensate in the total cost. With the adaptation of competitive structure of the power sector both in the generation and distribution levels, the transparent pricing structure of both the real and reactive power has become essential for wholesale competition to exist.

In the modern restructured power market, the pricing of active energy and other services is linked with the reactive power cost. Hence, it is assumed to be an ancillary service and priced separately. The fair pricing of these services may cause market liquidity, which may further result in approaching the optimal condition

Limit of armature current: I^2R power loss is caused due to the heavy armature current. This energy loss must be eliminated to limit the temperature rise in the conductor. The limit of armature current may be defined as the maximum value of the armature current that an armature may sustain without exceeding the heating boundary [25]. Heating limit of armature current is shown in Figure 5.1:

$$S = P + jQ = E \cdot I \quad (5.1)$$

$$P = E_t I_t \cos\emptyset \quad (5.2)$$

$$Q = E_t I_t \sin\emptyset \quad (5.3)$$

Limit of field current: Due to the heat results from $R_{fd} \, I_{fd}^2$ power loss, field current causes imposing of the second limit throughout the operation of the generator. With the help of steady-state equivalent circuit, the constant locus can be developed and plotted.

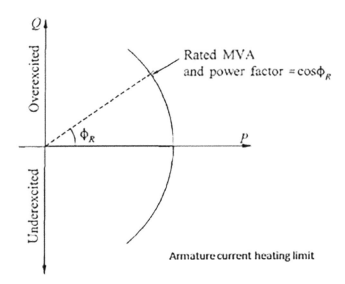

FIGURE 5.1 Heating limit of armature current.

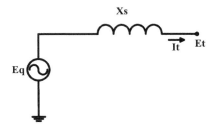

FIGURE 5.2 Cylindrical pole rotor generator steady-state model.

Cylindrical pole rotor generator: With $X_d = X_q = X_s$, the equivalent circuit of Figure 5.2 gives the relationship between E_t, I_t, and E_q. The phasor diagram of cylindrical pole rotor generator with Ra neglected is shown in Figure 5.2.

$$\hat{E}_q = \hat{E}_t + j\hat{I}_t X_s \tag{5.4}$$

$$|E_q| = X_{ad} i_{fd} \tag{5.5}$$

Solving equation for $\hat{E}_t$,

$$(X_{ad} i_{fd}) \sin \delta_i = (X_t I_t) \cos \varnothing \tag{5.6}$$

$$(X_{ad} i_{fd}) \cos \delta_i = E_t + (X_t I_t) \cos \varnothing \tag{5.7}$$

Rearranging yields (Figure 5.3)

$$I_t \cos \varnothing = \frac{(X_{ad} i_{fd}) \sin \delta_i}{X_s} \tag{5.8}$$

$$I_t \sin \varnothing = \frac{(X_{ad} i_{fd}) \cos \delta_i - E_t}{X_s} \tag{5.9}$$

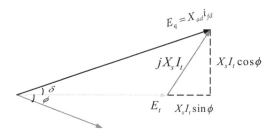

FIGURE 5.3 Cylindrical pole rotor generator steady-state phasor diagram.

Reactive Power Management in Power Systems

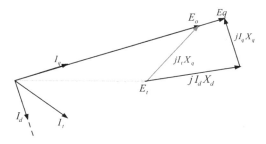

FIGURE 5.4 Salient pole rotor generator steady-state phasor diagram.

Therefore,

$$P = \frac{X_{ad}}{X_s} \cdot |E_t||I_t|\sin\delta \tag{5.10}$$

$$Q = \frac{X_{ad}}{X_s} \cdot |E_t||I_t|\cos\delta - \frac{|Et|^2}{X_s} \tag{5.11}$$

It can be concluded that for a given field current, the relationship between active and reactive powers is a circle which is centered at $-\frac{E_t^2}{X_s}$ on the Q axis and with $\left(\frac{X_{ad}}{X_s}\right)E_t i_{fd}$ as the radius (Figure 5.4).

5.10.1 Salient Pole Rotor Synchronous Generator

$$E_o = E_t + jI_t X_q \tag{5.12}$$

$$E_q = |E_o| + I_d(X_d - X_q) \tag{5.13}$$

$$P = \frac{|E_t||E_o|}{X_d}\sin\delta + \frac{|Et|^2}{2}\left(\frac{1}{X_q} - \frac{1}{X_d}\right) \tag{5.14}$$

$$Q = \frac{|E_t||E_o|}{X_d}\cos\delta - \frac{|Et|^2}{2}\left(\frac{1}{X_q} - \frac{1}{X_d}\right)\cos 2\delta - \frac{|Et|^2}{2} \tag{5.15}$$

Maximum reactive power generation limit will be maximum when $\delta = 0$; hence,

$$Q_{max} = \frac{|E_t||E_o|}{X_d} - \frac{|Et|^2}{2}\left(\frac{1}{X_q} - \frac{1}{X_d}\right) - \frac{|Et|^2}{2} \tag{5.16}$$

It may be concluded that the possible effect of the maximum field current rating on the capability of the machine may be drawn on the P-Q plan as shown in Figure 5.5.

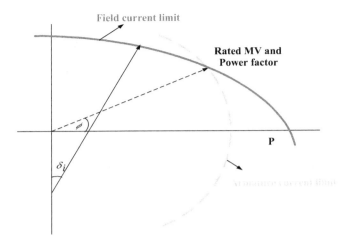

FIGURE 5.5 Heating limit of field current.

Limit of end region heating: During the operation of synchronous machine, the third limit is imposed by the local heating in the limit of end region [41]. For a particular type of excitation, the operating capacity of the machine gets affected by the heating limit of the end region [42].

The region of heating with armature current limit and field current limit is shown in Figure 5.5, in which the region of end turn for a generator is presented. Also, the end ring causes leakage flux and then enters and leaves in a direction which is axial perpendicular to stator laminations and causes initiation of eddy current in the laminations. This process further causes localization of heating in the area of end region.

During the overexcited mode of operation, for end leakage flux to be small, the high value of the field current helps in maintaining the ring saturated. But during the under-excited mode of operation, the value of this field current is very small, and hence, the ring will not be saturated at all. The armature current and field current are increased and added to the flux [41].

So, from the above conclusion, it may be concluded that in the end region, axial flux is increased by end turn flux. In the case of round rotor machines, this excessive heating the end ring may adversely affect the generator output [40].

5.11 SYNCHRONOUS MACHINE AS A SYNCHRONOUS CONDENSER

In the present restructured electricity market modal, the structure for the calculation of pricing of reactive power using hydroelectric generating units for condenser mode of operation is under development considering the variables associated with the system. Considering the case when support of reactive power is specially imposed by some hydroelectric unit using condenser mode of operation, a proposed structure developed and discussed in [39] is implemented in this article. In this structure, the contribution of reactive power support by thermal power plant is not taken into

Reactive Power Management in Power Systems

consideration. Koyna Hydro Electric Power Plant (KHEPP) [40,41] is the example where the methodology pricing of reactive power using hydroelectric plant is applied in condenser mode of operation.

It is seen that in synchronous mode of operation, the unit of hydroelectric may also be used. In this mode of operation, system with increased stability of the system and improved power factor is achieved. The machine must be operated in synchronous generator mode if it is started using condenser mode of operation from standstill condition till the synchronization with the grid. The changeover must be carried out to the condenser mode in the synchronization with grid mode only.

In condenser mode of operation, either the turbine is separated with the help of clutch, which is placed in between turbine and generator, or the turbine may remain connected to the generator, which acts in the form of load.

In the process of changeover, the net reduction in the power flow to the turbine causes the excitation of synchronous machine to reduce gradually. This causes decrement in the induced voltage E, with respect to the terminal voltage V of the synchronous generator. This causes the flow of power, from grid to the synchronous machine which further causes stopping of water flow to the turbine. The process ends when the machine does not generate active power.

During this time, changeover takes place from generator to condenser mode, causing the machine to be operated either in overexcited or under-excited mode. The machine is said to be in synchronous condenser or synchronous reactor mode, if there is no supply of reactive power. In either of the excited mode, i.e., overexcited or under-excited, the reactive power can be controlled by varying the excitation current.

For the generator and condenser mode as shown in Figure 5.6, the loading capability curve can be obtained using the characteristic equations of a synchronous machine with four regions of operation. This curve shows the relationship between reactive power and active power for different power factor conditions.

It is clear that condenser mode of operation region is smaller as compared to generator mode of operation. The indication of negative (−) power is that the active power is taken back from the grid for the machine to operate in condenser mode as well as for supplying additional loss.

Reactive power is fed to the system in condenser and generator mode, while absorption of reactive power takes place in Regions II and IV. Regions III and IV have been considered for further analysis of the system. An overexcited synchronous condenser in Region III works in leading power factor for reactive power supply, whereas in under-excited synchronous condenser mode works in lagging power factor to absorb reactive power. In Regions III and IV, the curve line shows that when reactive power flow support increases, the active power flow as an input to the machine also increases because of higher losses (Figure 5.6).

When the machine is operating in the generating mode, sometimes it is not capable of extracting more reactive power beyond the prescribed limit. The main advantage of this mode is that the maximum amount of reactive power that can be extracted is equal to one third of rated active power available. In comparison to the condenser mode of operation, the total amount of extracted reactive power lies in the range of 85%–95% of the rated capacity of the generator. This amount of power can

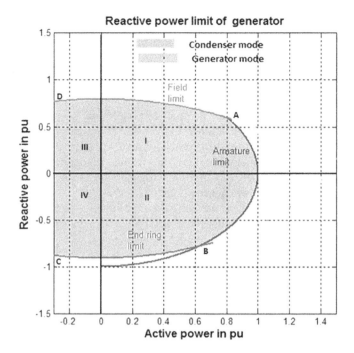

FIGURE 5.6 Distinct regions of generator and condenser mode.

be regulated quickly and effectively adjusted. Using overexcited and under-excited regions of operation, the limits of reactive power are higher in condenser mode as compared to generating mode.

5.12 CONCLUSION

In the modern power system generation, transmission, and application, there is a serious concern of reactive power measurement, modification, and implementation. In the present scenario where there is a huge and abrupt demand of electricity from the customer side, the generating companies are trying hard to meet the demand expectation and to provide quality power to the user. This quality power not only ensures the reliable operation of user's equipment but also provides the safeguard to the working personals. Hence, reactive power management for any interconnected system becomes important and customary in the present situation of fluctuating power availability. Apart from this, the operating conditions of the system should also not be deviated from the point values.

As soon as the operating conditions of the system change, the overall behavior of the integrated interconnected system also changes. In this regard, the analysis of active power, reactive power, and real power with changed condition must be analyzed to avoid system breakout. Proper online/offline monitoring of current, voltage, and several important parameters can be performed and may be executed for system restoration in case of failure. But even if the system violates the stability

Reactive Power Management in Power Systems

limits, the evaluation of sensitivities comes into picture for determining very fine and unpredictable changes in the system. Several types of sensitivity analysis and their outcome are presented in the abovementioned methods and procedures. Once the sensitivities are calculated, the exact and possible cause of system failure can be estimated, and the region of stability for the large integrated interconnected system can be predicted.

The case study and example of IEEE 14-bus integrated system are explained showing the effect of sensitivities on the calculation of reactive power management. FACTS devices are also presented with their new and updated model structure for maintaining the power flow for the system under study. It is seen that with the help of FACTS devices, the reactive power may regulate the line parameters, specially the voltage levels to supply uninterrupted active power flow, thereby increasing the overall efficiency of the interconnected system.

REFERENCES

1. Riesz J., Gilmore J. and MacGill I. 2015. Frequency control ancillary service market design: insights from the Australian national electricity market. *The Electricity Journal (Elsevier)* 28: 86–99.
2. Hirst E. and Kirby B. 1996. Costs for electric-power ancillary services. *The Electricity Journal*, 9: 26–30.
3. Singh S.N., Parida S.K., and Shrivastava S.C. 2009. Ancillary sevices management policies in India, An overview and key issues. *The Electricity Journal*, 22: 88–97.
4. Rebours Y.G., Kirschen D.S., and Trotignon M. 2007. A survey of frequency and voltage control ancillary services—Part I: Technical features. *IEEE Transactions on Power Systems*, 22: 350–357.
5. Singh S.N., Parida S.K., and Shrivastava S.C. 2009. Reactive power cost allocation by using a value-based approach. *IET Generation Transmission Distribution*, 3: 872–884.
6. Bhattacharya K. and Zhong J. 2001. Reactive power as an ancillary service. *IEEE Transactions on Power Systems*, 16: 294–300.
7. The National Grid Company.1999. NGC Reactive Market Report.
8. Ahmed S. and Strbac G. 1999. A method for simulation and analysis of reactive power market. *Proceedings of Power Industry Computer Applications Conference*, 13:337–341.
9. New York Independent System Operator. 1999. NYISO Ancillary Services Manual, 1999.
10. Shangyou H., and Papalexopoulos A. 1997. Reactive power pricing and management. *IEEE Transactions on Power Systems*, 12: 95–102.
11. Maxwell M., and Hawary E.I. 1999. A summary of algorithms in reactive power pricing. *International Journal Electric Power Energy Systems*, 21:119–24.
12. Horacio J., Jimenez-Guzman M., and Gutierrez-Alcaraz G. 2005. Ancillary reactive power service allocation cost in deregulated markets: A methodology. *International Journal Electric Power Energy Systems*, 27: 371–378.
13. Dai Y., Ni Y.X., and Shen C.M.2001. A study of reactive power marginal price in electricity market. *International Journal Electric Power Energy System*, 57: 41–48.
14. Zhang T., Elkasrawy A., and Venkatesh B. 2009. A new computational method for reactive power market clearing. *International Journal Electric Power Energy Systems*, 31: 285–293.
15. Alturky Y.A., Lo K.L. 2010 Real and reactive power loss allocation in pool-based electricity markets. *International Journal Electric Power Energy Systems*, 32: 262–270.

16. Lamont J.W. and Jian F. 1999. Cost analysis of reactive power support. *IEEE Transactions on Power Systems*, 14: 890–895.
17. Miller T.J. 1982. *Reactive Power Control in Electric Systems*. Hoboken, NJ: Wiley.
18. Ranatunga R.A.S.K., Annakkage U.D., and Kumble C.S. 2003.Algorithms for incorporating reactive power into market dispatch. *International Journal Electric Power Energy Systems*, 65: 179–186.
19. Chung C.Y., Chung T.S., Yu C.W., and Lin X.J. 2004. Cost-based reactive power pricing with voltage security consideration in restructured power systems. *International Journal Electric Power Systems* 70: 85–91.
20. Tiwari A., and Ajjarapu V. 2007. Reactive power cost allocation based on modified power flow tracing methodology. In *IEEE Power Engineering Society General Meeting*, pp. 1–7.
21. Vaidya G.A., Gopalakrishnan N., and Nerkar Y.P. 2011. Reactive power pricing structure for hydroelectric power station in condenser mode operation. *International Journal Electric Power Energy Systems*, 33: 1420–1428.
22. Chattopadhyay D., Bhattacharya K., and Parikh J. 1995. Optimal reactive power planning and its spot pricing: An integrated approach. *IEEE Transaction on Power Systems*, 10: 2014–2019.
23. Kumar J., and Kumar A. 2011. ACPTDF for multi-transactions and ATC determination in deregulated markets. *International Journal of Electrical and Computer Engineering (IJECE)* 1: 71–84.
24. Sawhney H., and Jeyasurya B. 2004. Application of unified power flow controller for available transfer capability enhancement. *Electric Power Systems Research*, 69:155–160.
25. Farahmand H., Rashidi-Nejad M., and Fotuhi-Firoozabad M. 2004. Implementation of FACTS devices for ATC enhancement using RPF technique. In *Proceedings of the IEEE, Power Engineering*, pp. 30–35.
26. Menniti D., Scordino N., and Sorrentino N. 2006. A new method for SSSC optimal location to improve power system available transfer capability. In *Proceedings of the IEEE, Power Systems Conference and Exposition*, pp. 938–945.
27. Reactive power support services in electricity markets.2001. Power System Engineering Research Center (PSERC) publication.
28. Kumar S., Kumar A., and Sharma N.K. 2013. Role of PMU & FACTS Controllers in Voltage Stability Analysis of Integrated Wind Farms. RTESC, NIT Kurukshetra, pp. 34–38.
29. Paucar V.L. and Rider M.J. 2001. Reactive power pricing in deregulated electrical markets using methodology based on theory of marginal costs. In *IEEE PES General Meeting*.
30. Cooper Crouse Hinde. 1999. Hydrogen cooled alternator purging. Hitech Instruments Report.
31. Baudry R.A., and King E.I. 1965. Improved cooling for generators of large rating. Power apparatus and systems. *IEEE Transaction on Power Systems*, 84: 106–114.
32. Ross M.D.1931. The application of hydrogen cooling to turbine generators. *American Institute of Electrical Engineer Transactions* 50: 381–386.
33. Nilsson N.E., and Mercurio J. 1994. Sysnchronous generator capability curve testing and evaluation. *IEEE Transactions on Power Delivery*, 9: 414–424.
34. Greene S., Dobson I., and Alvarado F.L. 2002. Sensitivity of transfer capability margin with a fast formula. *IEEE Transaction on Power Systems*, 17: 34–40.
35. Kumar S., Kumar A., and Sharma N.K. 2017. Sensitivity analysis based performance and economic operation of wind integrated system with FACTS devices for optimum load dispatch. *Renewables-Wind, Water and Solar*, 4: 1–13. Springer.
36. Ghawghawe N.D. and Thakre K.L. 2009. Computation of TCSC reactance and suggesting criterion of its location for ATC enhancement. *International Journal on Electric Power and Energy Systems*, 31: 86–93.

37. Kumar S., and Kumar A. 2020. Design and optimization of multiple FACTS devices for congestion mitigation using sensitivity factor with wind integrated system. *IETE Journal of Research*. doi:10.1080/03772063.2020.1787872.
38. Xiao Y., Song Y.H., and Sun Y.Z. 2003. Available transfer capability enhancement using FACTS devices. *IEEE Transactions on Power Systems*, 18: 305–312.
39. Hasanpour S., Ghazi R., and Javid M.H. 2001. A new approach for accurate pricing of reactive power and its application to cost allocation in deregulated electricity markets. *Journal of Iranian Association of Electrical and Electronics Engineers*, 7: 31–39. Fall & Winter.
40. Schweppe F.C., Caramanis M.C., and Bohn R.E. 1999. *Locational based Pricing of Electricity*. Kluwer Academic Publishers.
41. Baughman M.L., and Siddiqi S.N. 1993.Real time pricing of reactive power: Theory and case study results. *IEEE Transactions on Power Systems*, 6: 23–29.
42. Dandachi N., Rawlins M., O., and Stott B., 1996.OPF for reactive power transmission pricing on the NGC system. *IEEE Transaction on Power Systems*, 11: 226–232.

6 Optimum Scheduling and Dispatch of Power Systems with Renewable Integration

Abhishek Rajan and Bimal Kumar Dora
National Institute of Technology Sikkim

CONTENTS

6.1 Introduction 132
    6.1.1 Optimal Reactive Power Dispatch 132
6.2 Related Work 134
6.3 Problem Formulation 135
    6.3.1 Objective Functions 137
        6.3.1.1 Minimization of Active Power Loss 137
        6.3.1.2 Minimization of TVD 137
        6.3.1.3 Minimization of Maximum L-Index 137
        6.3.1.4 Formulation of Overall Objective Function 138
    6.3.2 Constraints 139
        6.3.2.1 Equality Constraints (ECs) 139
        6.3.2.2 Inequality Constraints (ICs) 139
    6.3.3 Constraints Handling 140
        6.3.3.1 Handling ECs 140
        6.3.3.2 Handling ICs 141
6.4 Comparative Results on Optimal Reactive Power Dispatch 141
6.5 Optimal Active Power Dispatch 149
    6.5.1 Problem Formulation of OAPD problems 150
6.6 Optimal Reactive Power Dispatch with Renewable Power Generations 153
    6.6.1 Classification of Wind Turbines 155
6.7 Active and Reactive Power Sharing in Microgrid 158
6.8 Conclusion 159
References 159

DOI: 10.1201/9781003271857-6

6.1 INTRODUCTION

6.1.1 Optimal Reactive Power Dispatch

The power system is considered one of the most nonlinear complex systems in science and engineering. It is possibly the largest manmade network that connects the village to city, city to state, state to country, and country to another country. In the early phase of the power system, researchers and practitioners were much concerned about the operational cost of the power system. Ample research has been done to optimize the cost of the production as well as the distribution. Gradually, with time progress, the power system experienced exponential growth in nonlinear loads from industries and households. Due to the continuous increase in the nonlinear load demands, the planning and allocation of reactive power became the primary concern of power system utilities and researchers. Since there exists a strong correlation between reactive power and system bus voltage magnitude, any imbalance or improper planning of reactive power may give rise to voltage instability or voltage collapse in the worst case (Wood and Wollenberg 1996). Voltage instability has a very adverse effect on the economy, security, line loadings, and life of the equipment connected to the power system. A loud change in the magnitude of the voltage sometimes may also result in insulation failure. Hence, in this era of industrialization and a tremendously growing population, proper planning, management, and reactive power allocation in the power system are significant areas of concern for power system practitioners and researchers. Continuous efforts are being made to maintain the system voltage within their safe limits by properly utilizing and controlling the power system's voltage-controlled devices. Simultaneous control of these voltage-controlled devices for proper distribution of reactive power in the entire power system is complex and challenging. It requires an appropriate mechanism that can guide the power system operators to adjust voltage-controlled devices' outputs. In the power system, this mechanism refers to optimal reactive power dispatch (ORPD) (Carpentier 1979). Optimal power flow (OPF) is a power system optimization tool that ensures a stable power system operating state while optimizing a specific objective, satisfying all the physical and operational constraints in the power system (Carpentier 1979). In particular, ORPD optimizes the voltage-related objective functions such as active power losses, voltage deviation, and voltage stability indexes. While optimizing any of these objectives, it also takes care of the various operational constraints like load voltage magnitude, slack bus power, reactive power output limit of generators, and line flows. ORPD was first introduced by Carpentier in 1962 (Carpentier 1979).

It is nearly impossible to manually find the optimal voltage control devices' settings (Dommel and Tinney 1968). Hence, it requires an efficient computation tool and robust algorithm to perform the task. Large numbers of classical optimization techniques have been explored to solve ORPD problems by researchers (Lee et al. 1985, Bjelogrlic et al. 1990, Granville 1994). Classical Optimization techniques being exact methods provide excellent convergence characteristics and accurate results for those problems whose objective functions and constraints are linear, continuous, and differentiable. However, they sometimes fail to optimize for those functions and constraints that are nonlinear, nondifferentiable, and discontinuous. In particular, it can be said that the presence of nonconvexity and discontinuity in objective functions and constraints in ORPD problems made the solution computationally complex and challenging for classical optimization techniques (Granville 1994). To overcome the drawback of classical optimization techniques, population-based approaches have been developed that mimic natural species' characteristics or the physical phenomenon (Goldberg 1989). Some of the famous population-based algorithms are genetic algorithm (GA), ant colony optimization (ACO), particle swarm optimization (PSO), artificial bee colony (ABC), firefly algorithm (FA), harmony search algorithm (HAS), bat algorithm (BA), gravitational search algorithm (GSA), chemical reaction algorithm (CRA), black hole-based optimization (BHBO), water cycle optimization (WCO), teaching learning-based optimization (TLBO), exchange market algorithm (EMA), etc. GA is considered as the first algorithm based on the evolution of human beings. It was first observed by John Holland in the early sixties and proposed his theory of GA based on Darwin's theory of evolution. In 1989, the student of J Holland, David E. Goldberg, extended his theory and gave a complete mathematical shape to GA (Goldberg 2002). GA tries to mimic the biological reproduction and selection process to find the fittest solution (Goldberg 2002). This algorithm was far more efficient than classical optimization techniques like random search and exhaustive search algorithms. However, they do not require any extra information about the given problem (Kinnear 1994). Algorithms like ACO, PSO, ABC, and FA are nature-inspired algorithms based on natural creators' swarm intelligence for food foraging. They perform socially as well as local searches to find their food sources. In 1992, Marco Dorigo, working on optimization and natural algorithms (Dorigo 1992), presented his work on ACO. The search technique of this algorithm was inspired by the swarm intelligence of social ants, which uses the pheromone as a chemical messenger. Later in 1995, James Kennedy and Russell C. Eberhart (1995) came up with an algorithm called PSO. This algorithm was more promising and considered a significant contribution in optimization and natural algorithms. Loosely speaking, PSO is inspired by the swarm intelligence of natural species. All the swarms start with an initial random position and search for the food. During their search, they share information to focus on quality solutions. Since its development, there is strong evidence suggesting that PSO and its variants have been utilized in different science and engineering problems and produce some outstanding results compared with traditional algorithms. In some problems, it also outperforms GA and ACO though not conclusive. In the year 1996, Storn and Price developed a new vector-based algorithm called differential evolution (DE) (Storn and Price 1997). This was the time when Wolpert and Macready came up with a theory called *No Free Lunch Theory* in 1996 (Wolpert and Macready 1997). The theory says that "if algorithm "A" performs better than algorithm "B" for some optimization functions, then "B" will outperform "A" for other functions. That is to say, if averaged overall possible function space, both algorithms "A" and "B" will perform, on average, equally well." *Alternatively, no universally better algorithms exist* (Wolpert and Macready 1997). As all real-world problems are different, and generally, we require the exact solution of every problem, we need an algorithm that works explicitly well in solving the particular problem. Average solutions from different algorithms may not work due to the difficulties of choosing a specific solution from many solutions. Therefore designing and modification of the multipurpose algorithms is still an open area of research.

Many algorithms have been developed in view of the above, and many are modified to solve several real-life optimization problems. Many of them have proved their ability to solve highly complex and challenging issues in the diverse field of science and engineering. Electrical engineering, one of the most complex branches, possesses a vast research domain related to highly nonlinear and constrained optimization problems. ORPD is one such example of a highly nonlinear, nonconvex constrained optimization problem related to power systems. Many researchers have utilized a different meta-heuristic algorithm to solve ORPD. Results obtained from some of the algorithms were excellent, which further motivated the researchers to apply more efficient algorithms to extract some convincing results under a tough, constrained environment.

6.2 RELATED WORK

As discussed in the previous section, ORPD is a high-degree nonlinear and optimization problem of a power system that includes several level constraints to satisfy while optimizing specific objectives. The control parameters of ORPD problems are diverse (some are continuous and some discrete), which sometimes transform the optimization problem into a nonlinear optimization problem with mixed integers (MINLP). Hence, it appears to be very challenging for any optimization technique to efficiently find the optimal solution of ORPD problems under the said complex constrained environment. In view of the above, several population-based algorithms have been tested on solving ORPD under relaxed and constrained environments. Some of them are discussed below:

Subbaraj and Rajnarayanan have implemented GA (Subbaraj and Rajnarayanan 2009) to solve ORPD problems. The main motive was to lessen system power losses. Dai et al. (2009) utilize seeker optimization algorithm to solve the same. PSO with comprehensive learning (CLPSO) was developed by Mahadevan and Kannan (2010). The results are tested on standard IEEE 30- and 118-bus systems to minimize the active power loss, improve the voltage profile, and improve the voltage stability index. Duman et al. (2011, 2012a) designed an algorithm based on GSA's physical phenomenon. They implemented the same to solve ORPD problems and optimize the same objectives as discussed in (Mahadevan and Kannan, 2010). The results obtained are compared with several other state-of-the-art and found that the said algorithm performs well even for the bigger systems. Similarly, many more algorithms like DE (Sayah and Zehar 2008, Ela et al. 2011), Biogeography-Based Optimization (BBO) (Bhattacharya and Chattopadhyay 2011, 2010), HSA (Sivasubramani and Swarup 2011), GSA (Shaw et al. 2014), cuckoo search algorithm (CSA) (Biswas et al. 2014), BHBO (Bouchekara 2014), EMA (Rajan and Malakar 2016), ant lion optimization (ALO) (Rajan et al. 2017), water cycle algorithm (WCA) (Eskandar et al. 2012), etc. have been implemented to solve ORPD problems. The above-discussed algorithms are nature-inspired, and they have population-based search techniques, unlike classical optimization methods, which generally have only one initial point. Most classical optimization techniques utilize differential calculus to find the optimal value of the given objective function. Whereas population-based optimization techniques are free from any predefined optimization method, they enjoy the random searching capability over a defined search space. They are flexible and try to maintain their exploration and exploitation while searching for the optimal solution.

On the contrary, it has also been observed that any imbalance between exploration and exploitation may results in poor performance of the algorithm like premature convergence, trapping into local optima, and extensive simulation time (Sayah and Zehar 2008, Ela et al. 2011, Bhattacharya and Chattopadhyay 2011, 2010, Sivasubramani and Swarup 2011, Shaw et al. 2014, Biswas et al. 2014, Bouchekara 2014, Rajan and Malakar 2016, Rajan et al. 2017). Since the search technique is a random process, single-time execution is insufficient to declare the solution as an optimal one. Several trial runs are required to justify the optimality of the solution. Keeping the above facts in mind, many researchers have tried to solve these problems by enhancing the search capability of algorithms either by modifying their potential of exploration and exploitation or by hybridizing it with other intelligent methods. It is seen from the literature that the modification over the internal structure of original algorithms enhances their search capabilities. Some important literature is discussed below. As an application, only those modified/hybrid algorithms are discussed which are used to solve ORPD. In (Rajan and Malakar 2015, Panigrahi and Pandi 2008, Khorsandi et al. 2010), FA, bacterial foraging optimization (BFO), and Shuffled Frog Leaping Algorithm (SFLA) has been hybridized using a classical optimization technique called Nelder-Maid Simplex (NMS). Classical optimization techniques work exceptionally well on functions with single optima. Hence, they can help meta-heuristic algorithms in escaping out from the problem of trapping into local optima. In all the literature mentioned above, NMS is called a subroutine when the original algorithm appears to be trapped in local optima. Ding combined CSA with PSO to improve the search direction and convergence characteristics. Fatah et al. (2019) modified the original sine cosine algorithm (SCA) steps to bring a proper balance between exploration and exploitation of the algorithm. Similarly, Jangir et al. (2016) combined multi-verse optimizer (MVO) with PSO to improve the search capability of the original algorithms. The results demonstrate that PSO-MOV has a fast convergence rate and outperforms both the original methods. Ghasemi et al. (2014b) modified the TLBA using the double DE algorithm (DDEA) to improve the convergence characteristics of TLBA and utilize it to minimize active power loss. The performance of the modified algorithm is checked on small, medium, and extensive systems. The efficiency and robustness of the algorithm are also weighted by performing statistical analysis. All the above-said hybrid/improved algorithms are utilized to solve ORPD. The results obtained are also compared with contemporary algorithms, and it is observed that they outperform many of them.

6.3 PROBLEM FORMULATION

ORPD is an optimization problem, and it is mathematically formulated as follows (Wood and Wollenberg 1996, Carpentier 1979):

$$\text{Minimize} \quad f(x, u) \tag{6.1}$$

$$\text{Subjected to} \quad \left\{ \begin{array}{l} g(x,u) = 0 \\ h_{\min} \le h(x,u) \le h_{\max} \end{array} \right. \tag{6.2}$$

where f is a optimization function, and x u are the vectors of state and control variables, respectively. $g(x, u)$ and $h(x, u)$ are the equality and inequality constraints based on state and control variables. Since ORPD problem deals with the optimal management of reactive power in the system so as to maintain the flat voltage profile under steady-state operation, control or independent variables (u) in this type of problems are outputs of those components whose adjustment can manage the reactive power in the system. In power system, there are three components which predominantly regulate the reactive power flow. These are as follows:

1. Output voltage of alternators/generators (V_G) (which can be changed by adjusting filed supply)
2. Position of online transformers, tap changing type (Tap).
3. VAR outputs of the compensators (Q_C).

Therefore, the vector of control variable (u) can be represented mathematically as

$$u^T = \left[\overbrace{V_{G1}, \ \dots V_{G_{\text{NPV}}}}^{\text{continuous}} \ \overbrace{\text{Tap}_1, \dots \text{Tap}_{\text{NT}}, \ Q_{C1}, \dots Q_{C_{\text{NC}}}}^{\text{discrete}} \right] \tag{6.3}$$

where N_{PV} represents the number of photo-voltaic (PV) buses in the system. NT and NC are the number of transformers (tap changing type) and VAr compensators present in the system. It is to be noted that active power output of generators can also be adjusted by adjusting the fuel input. Since there exists a very weak correlation between active power (P) and voltage (V), P is kept fixed during the execution of ORPD problems. The vector of dependent variable (x) is mathematically represented as

$$x^T = \left[P_{G1}, \ V_{L1}, \dots V_{L_{\text{NPQ}}}, \ Q_{G1}, \ Q_{G2}, \dots Q_{G_{\text{NPV}}}, \ S_{L1}, \dots S_{L_{\text{NTL}}} \right] \tag{6.4}$$

where (P_{G1}) and $V_{L1}, \dots V_{L_{\text{NPQ}}}$ are power output of slack generator and voltage of PQ buses, respectively. Load buses are also called PQ buses because of the fact the active and reactive power consumption of the load is specified at these buses. The other dependent variables are Q output of generators $(Q_{G1}, Q_{G2}, \dots Q_{G_{\text{NPV}}})$ and S in the transmission line $(S_{L1}, \dots S_{L_{\text{NTL}}})$. The variables in Eq. (6.4) are called dependent variables because the outputs of these variables change with the adjustment of control variable (u) (6.3). The solution of ORPD problem is said to be feasible if and only if the outputs of dependent variable (6.4) will be well within their limit for that solution. This is one of the major challenge meta-heuristic algorithms are facing while optimizing certain objective of ORPD problem.

Optimum Scheduling with Renewable Integration

6.3.1 Objective Functions

Generally, three objectives are considered while solving the ORPD problems. These objectives are (1) minimization of active power loss, (2) minimization of total voltage deviation (TVD), and (3) minimization of maximum L-index.

6.3.1.1 Minimization of Active Power Loss (Duman et al. 2012a, Ghasemi et al. 2014b)

The active power loss in the transmission line is mathematically expressed as

$$P_{loss} = \sum_{\substack{k \in NTL \\ k=(i,\,j),\; i \neq j}} G_k \left(V_i^2 - V_j^2 - 2V_i V_j \cos\theta_{ij} \right) \tag{6.5}$$

where V_i and V_j are the node voltages of ith and jth buses, respectively. θ_{ij} is the load angle between both the buses. G_k is the conductance of kth transmission line.

6.3.1.2 Minimization of TVD (Duman et al. 2012b, Ghasemi et al. 2014b)

Optimizing the total load voltage deviation in the power system helps the system to approximately maintain a flat voltage profile throughout the system. It further helps to operate the system more securely. TVD is treated as one the security indexes of the power system. Mathematically it is expressed as

$$TVD = \sum_{i=1}^{N_{PQ}} \left| V_i - V_i^{ref} \right| \tag{6.6}$$

$$V_i^{ref} = 1 \text{p.u.}$$

6.3.1.3 Minimization of Maximum L-Index (Duman et al. 2012b, Ghasemi et al. 2014b)

L-index is the voltage stability related security index of the power system network. It is also known as voltage stability index. Every load bus has a scalar value of L-index (L_j) which gets changed with the change in reactive power flow in the system. The maximum and minimum values of L_j are defined as 1 and 0, respectively. If L_js are close to 0, the system is said to be voltage stable and considered as highly unstable if it is close to 1. The high value of L_j is alarming as it indicates the situation of voltage collapse. Therefore, it is of utmost importance to keep the value of all L_js close to 0. In the optimization problem, the maximum value of L_j among all L_js is minimized, which automatically signifies that other L_js are less than the maximum L_j. Mathematically, it is expressed as

$$L_j = \left| 1 - \sum_{i=1}^{N_{PV}} F_{ij} \frac{V_i}{V_j} \right|, \; j \in N_{PQ} \tag{6.7}$$

$$L_{j\max} = \max\left(L_j \right)$$

$$\text{Min}\left(L_{j\text{max}}\right) \tag{6.8}$$

$$F_{ij} = -\left[Y_1\right]^{-1}\left[Y_2\right] \tag{6.9}$$

$[Y_1]$ and $[Y_2]$ can be formulated as follows:

$$\begin{bmatrix} I_{\text{PQ}} \\ I_{\text{PV}} \end{bmatrix} = \begin{bmatrix} Y_1 & Y_2 \\ Y_3 & Y_4 \end{bmatrix} \begin{bmatrix} V_{\text{PQ}} \\ V_{\text{PV}} \end{bmatrix} \tag{6.10}$$

6.3.1.4 Formulation of Overall Objective Function

When any of the abovementioned objective functions is considered for optimization, all the operational constraints are to be satisfied simultaneously. Mathematically, the overall objective function is can be expressed as

$$\text{minimize } J = J_{\text{obj}} + \sigma_G \left(P_{G1} - P_{G1}^{\text{limit}}\right)^2 + \sigma_{Ui} \left(V_{Li} - V_{Li}^{\text{limit}}\right)^2$$

$$+ \sigma_{QGi} \left(Q_{Gi} - Q_{Gi}^{\text{limit}}\right)^2 + \sigma_{TL} \left(S_L - S_L^{\text{limit}}\right)^2 \tag{6.11}$$

J_{obj} is the objective function considered for optimization. Since this work deals with the single objective optimization, only one objective function among the three should be considered at a time. $\left(P_{G1} - P_{G1}^{\text{limit}}\right)$, $\left(V_{Li} - V_{Li}^{\text{limit}}\right)$, $\left(Q_{Gi} - Q_{Gi}^{\text{limit}}\right)$, and $\left(S_L - S_L^{\text{limit}}\right)$ are the violations in the dependent variables. Violations are squared to keep the value positive. $\sigma_G, \sigma_{Ui}, \sigma_{QGi}$, and σ_{TL} are the penalty factor (PF) associated with the corresponding violations. The suitable value of PF is determined using trial and error method. It is problem dependent and may differ with dimensions and complexity of optimization problem (Mezura-Montes and Coello 2011). $P_{G1}^{\text{limit}}, V_{Li}^{\text{limit}}, Q_{Gi}^{\text{limit}}$, and S_L^{limit} are the limiting values of dependent variable. The values of above-discussed dependent variables should be kept within their limiting band. This is achieved by the following method:

$$P_{G1}^{\text{limit}} = \begin{cases} P_{G1}^{\text{min}} & \text{if } P_{G1} < P_{G1}^{\text{min}} \\ P_{G1}^{\text{max}} & \text{if } P_{G1} > P_{G1}^{\text{max}} \\ P_{G1} & \text{if } P_{G1}^{\text{min}} \le P_{G1} \le P_{G1}^{\text{max}} \end{cases} \tag{6.12}$$

$$V_{Li}^{\text{limit}} = \begin{cases} V_{Li}^{\text{min}} & \text{if } & V_{Li} < V_L^{\text{min}} \\ V_{Li}^{\text{max}} & \text{if } & V_{Li} > V_L^{\text{max}} \\ V_{Li} & \text{if } V_{Li}^{\text{min}} \le V_{Li} \le V_{Li}^{\text{max}} \end{cases} \tag{6.13}$$

$$Q_{Gi}^{\text{limit}} = \begin{cases} Q_{Gi}^{\text{min}} & \text{if } Q_{Gi} < Q_{Gi}^{\text{min}} \\ Q_{Gi}^{\text{max}} & \text{if } Q_{Gi} > V_{Gi}^{\text{max}} \\ Q_{Gi} & \text{if } Q_{Gi}^{\text{min}} \le Q_{Gi} \le Q_{Gi}^{\text{max}} \end{cases} \tag{6.14}$$

Optimum Scheduling with Renewable Integration

$$S_L^{\text{limit}} = \begin{cases} S_L & \text{if} \quad S_L \leq S_L^{\max} \\ S_L^{\max} & \text{if} \quad S_L > S_L^{\max} \end{cases} \tag{6.15}$$

In Eqs. (6.12), (6.13), and (6.14), both upper and lower limits of the variable are defined, whereas in Eq. (6.15), only the upper limit is defined. This is because of the fact that in transmission line, power can flow in both directions so researchers are more concerned about the amount of power flow not in the direction of flow.

6.3.2 Constraints

ORPD problem deals with several physical and operational constraints. It includes both equality and inequality constraints. The detailed information about the constraints used in this work are discussed below.

6.3.2.1 Equality Constraints (ECs)

$$g(x, u) = \begin{cases} P_{Gi} - P_{Di} - |V_i| \sum_{j=1}^{NB} |V_j| \left\{ G_{ij} \cos\left(\theta_i - \theta_j\right) + B_{ij} \sin\left(\theta_i - \theta_j\right) \right\} = 0 \\ Q_{Gi} - Q_{Di} - |V_i| \sum_{j=1}^{NB} |V_j| \left\{ G_{ij} \cos\left(\theta_i - \theta_j\right) - B_{ij} \sin\left(\theta_i - \theta_j\right) \right\} = 0 \end{cases} \tag{6.16}$$

Equation (6.16) is a power balance equation generally used in Newton-Raphson load flow (NRLF). It is to be noted that output at all PV buses is fixed except for slack bus. Since active power output of slack generator (P_{G1}) changes with change in the transmission line losses, P_{G1} is considered as a dependent variable.

6.3.2.2 Inequality Constraints (ICs)

In ORPD problem, ICs are both on independent and dependent variables. If any of these constraints is violated, the solution is considered to be infeasible. The set of ICs on both the variables is mathematically represented in Eqs. (6.17) and (6.18):

$$h_{\min} \leq h(u) \leq h_{\max} \begin{cases} \text{On independent variable} \\ V_{Gi}^{\min} \leq V_{Gi} \leq V_{Gi}^{\max} \quad i \in N_{\text{PV}} \\ \text{Tap}_k^{\min} \leq \text{Tap}_k \leq \text{Tap}_k^{\max} \quad k \in \text{NT} \\ Q_{Ci}^{\min} \leq Q_{Ci} \leq Q_{Ci}^{\max} \quad i \in \text{NC} \end{cases} \tag{6.17}$$

Out of the above three, generator voltage (V_{Gi}) is continuous in nature, while (Tap_k) and (Q_{Ci}) are discrete in nature.

$$h_{\min} \leq h(x) \leq h_{\max} \tag{6.18}$$

Equation (6.18) represents the set of ICs on dependent variable. Q_{Gi} and P_{G1} are the reactive power output of generator and slack bus power, respectively. Both are dependent variables (reasons are already explained in Section 3.2.1), and their values are determined by executing the load flow solution method. V_{Li} and S_L are the load voltage and power flow in transmission line, respectively. They are called security constraints and must be confined within their operational limit. Any violations in the dependent variable may be a serious threat to the equipment or to a complete power system as a whole.

6.3.3 Constraints Handling

Any real-world optimization problem is associated with constraints. Constraint handling is a serious issue for any optimization technique as any solution with a violation in the constraints cannot be treated as a feasible one. Therefore, efficient constraint handling techniques are needed to be developed to treat the infeasible solution to restrict them within their feasible boundary without affecting the diversity of the overall population (Mezura-Montes and Coello 2011). ORPD, as discussed earlier, is a highly nonlinear optimization problem that contains several inequality and equality constraints (ECs) and needs proper treatment of their constraints. The strategies used for handling the constraints in the present work are discussed in Sections 6.3.3.1 and 6.3.3.2.

6.3.3.1 Handling ECs

ECs are considered the toughest/hard constraints in optimization problems as they reduce the region of feasible space. This effectively slows down the optimization process. Therefore, the general method of treating the ECs is by converting them into inequality constraints using small threshold value (Runarsson and Yao 2000). Some other methods have also been proposed to overcome the difficulties of ECs. Ullah et al. (2012) proposed a local search technique to handle ECs. This technique first reaches the point on equality constraint of any solution and then starts exploring based on the constraints area. This method sometimes fails when the optimization problem consists of many variables or when the maximum number of variables is nonlinear. When it searches in constraint landscape, it uses numerical search methods to evaluate the values of variables. Another method for handling ECs is presented by Schoenauer and Michalewicz (1996). Geometrical crossover is used for exploration. Other than this, very recently, Guohua et al. proposed an efficient method to handle equality constraints known as equality constraints and variable reduction strategy (ECVRS). In this method, numbers of ECs and variables are reduced, thereby improving the efficiency of evolutionary algorithms. In the present ORPD problem, numbers of ECs are present, and they increase with the increase in the bus size. For each bus, two ECs are associated (described in Eq. (6.16)), out of which one is related to active power balance, and another is for reactive power balance. If the total number of buses present in the system is "N", then "2N" numbers of ECs should be satisfied. In power system practice, these ECs are considered to be taken care of by the load flow solution methods. In the present work, NRLF solution method is utilized to solve the power flow problem. It is to be noted that NRLF converges only when the ECs presented in (Eq. 6.16) are completely satisfied.

Optimum Scheduling with Renewable Integration

6.3.3.2 Handling ICs

ICs are imposed in the system to maintain the system parameters well within their limit. ICs can be seen both on independent and dependent variable sides. ICs on independent variables can be handled efficiently by randomly generating their value within the given limits. Generally, the independent side control variables present in the ORPD problems are generator output voltages, tap changers positions, and output of shunt capacitors. These randomly generated control variables are first passed to load flow program and to determine the dependent variables as these are dependent on control variables, they are likely to violate their corresponding limit. At this stage, we need an efficient constraint handling mechanism to treat the violation in dependent variables. Several such constraint handling mechanisms have been discussed by Duman et al. (2012b). The most common is the PF method (Eq. 6.11).

6.4 COMPARATIVE RESULTS ON OPTIMAL REACTIVE POWER DISPATCH

System information of IEEE 30 bus and 57 bus (refer Table 6.1) are shown by Lee et al. (1985) and Duman et al. (2012a). The comparative result with different optimization methods is presented in Table 6.2.

From the results, it can be seen that researchers have applied different algorithms to optimize the active power loss in IEEE 30-bus system, and they have got convincing results. Similarly, researchers have also utilized several meta-heuristic algorithms to optimize voltage deviations and L-index in IEEE 30 bus. It is important to note that comparison results shown in this chapter are simulated on the same system under the same operational and physical constraints. The compression results for other objective functions related to ORPD are shown in Tables 6.3 and 6.4.

Researches have also tried their algorithms in more complex and higher dimension systems like IEEE 57- and 118-bus systems. The comparison analysis of one of the objectives for IEEE 57-bus system are shown graphically (Rajan et al. 2017):

TABLE 6.1
System Information

	IEEE 30 (Lee et al. 1985)	IEEE 57 (Duman et al. 2012a)
Total gen. bus:	6	7
Total buses:	30	57
Total branch:	41	80
Total transformer	4	15
Total SVC	9	3
Total PQ bus:	24	50
Base active load:	283.4 MW	1250.8 MW
Base reactive generation:	108.922 MVAr	345.45 MVAr
Initial P loss:	5.811 MW	28.462 MW
Initial Q loss:	32.417 MVAr	-124.27 MVAr

TABLE 6.2

Results of P Loss Minimization (IEEE 30-Bus System)

Control Variables	Initial	PSO (Mahadevan and Kannan 2010)	CLPSO (Mahadevan and Kannan 2010)	DE (Ela, et al. 2011)	BBO (Bhattacharya and Chattopadhyay 2011)	BFOA (Rajan and Malakar 2015)	TLBO (Ghasemi et al. 2014b)	QOTLBO (Ghasemi et al. 2014b)	BA (Rajan et al. 2017)	CSA (Rajan et al. 2017)	ALO (Rajan et al. 2017)
Vg1	1.05	1.1	1.1	1.1	1.1	1.1	1.1	1.1	1.1	1.1	1.1
Vg2	1.04	1.1	1.1	1.0944	1.0944	1.0261	1.0936	1.0942	1.0545	1.093	1.0944
Vg5	1.01	1.0867	1.0795	1.0749	1.0749	1.0696	1.0738	1.0745	1.0757	1.0741	1.075
Vg8	1.01	1.1	1.1	1.0768	1.0749	1.1	1.0753	1.0765	1.0768	1.0955	1.0768
Vg11	1.05	1.1	1.1	1.0999	1.0999	1.1	1.0999	1.1	1.1	1.1	1.1
Vg13	1.05	1.1	1.1	1.0999	1.0999	1.1	1.1	1.0999	1.1	1.1	1.1
T6–9	1.078	0.9587	0.9154	1.0465	1.0435	0.98	0.05	0.05	1.05	0.9	0.98
T6–10	1.069	1.0543	0.9	0.9097	0.9011	0.94	0.05	0.05	0.91	0.91	0.98
T4–12	1.032	1.0024	0.9	0.9867	0.9824	1.05	0.05	0.05	0.99	0.9	0.99
T27–28	1.068	0.9755	0.9397	0.9689	0.9691	0.98	0.05	0.05	0.98	0.9	0.97
Qc10	0	4.2803	4.9265	5	4.9998	3.1	0.0457	0.0445	2.96	4.99	4.69
Qc12	0	5	5	5	4.987	4.6	0.05	0.05	1.43	4.49	4.08
Qc15	0	3.0288	5	5	4.9906	5	0.0286	0.0283	0.96	4.91	4.94
Qc17	0	4.0365	5	5	4.997	2.1	0.05	0.05	1.67	4.87	3.31
Qc20	0	2.6697	5	4.406	4.9901	3.7	0.0258	0.0256	2.8	2.53	3.42
Qc21	0	3.8894	5	5	4.9946	2.3	1.0251	1.0664	2.8	3.58	4.89
Qc23	0	0	5	2.8004	3.8753	1.9	0.9439	0.9	2.61	4.41	4.81
Qc24	0	3.5879	5	5	4.9867	2.3	0.9992	0.9949	3.95	4.95	3.73
Qc29	0	2.8415	5	2.5979	2.9089	0.1	0.9732	0.9714	4.46	0.61	3.54
Min Ploss	5.812	4.6282	4.5615	4.555	4.5511	4.623	4.5629	4.5594	4.5786	4.5359	4.5372

(Continued)

Optimum Scheduling with Renewable Integration

TABLE 6.2 (Continued)
Results of P Loss Minimization (IEEE 30-Bus System)

Control Variables	MALO (Rajan et al. 2017)	GSA-CSS (Huang et al. 2012)	IGSA-CSS (Huang et al. 2012)	ABC (Rajan and Malakar 2015)	FA (Rajan and Malakar 2015)	HFA (Rajan and Malakar 2015)	GSA (Duman et al. 2012)	OGSA (Shaw et al. 2014)	EMA (Rajan and Malakar 2016)
Vg1	1.1	1.080987	1.081281	1.1	1.1	1.1	1.07165	1.05	1.1
Vg2	1.0943	1.071832	1.072177	1.0615	1.0644	1.054	1.02219	1.041	1.0942
Vg5	1.0752	1.049583	1.050142	1.0711	1.0745	1.07	1.04	1.0154	1.0749
Vg8	1.0972	1.049744	1.050234	1.0849	1.0869	1.08	1.05072	1.0267	1.0765
Vg11	1.1	1.087238	1.1	1.1	0.94	1.1	0.97712	1.0082	1.1
Vg13	1.1	1.07333	1.068826	1.0665	0.94	1.1	0.96765	1.05	1.1
T6–9	0.92	1.04	1.08	0.97	1	0.98	1.0984	1.0585	1.0311
T6–10	0.9	0.964	0.902	1.05	0.94	0.95	0.9824	0.9089	0.9
T4–12	0.91	1.02	0.99	0.99	1	0.97	1.0959	1.0141	0.9763
T27–28	0.93	0.972	0.976	0.99	0.97	0.97	1.05933	1.0182	0.9616
Qc10	5	0.0255	0	5	3	4.7	1.6537	3.3	4.6908
Qc12	5	0.0335	0	5	4	4.7	4.3722	2.49	5
Qc15	5	0.0315	0.038	5	3.3	4.7	0.1199	1.77	5
Qc17	5	0.035	0.49	5	3.5	2.3	2.0876	5	5
Qc20	5	0.026	0.0395	4.1	3.9	4.8	0.3577	3.34	4.9934
Qc21	5	0.03	0.05	3.3	3.2	4.9	0.2602	4.03	5
Qc23	5	0.035	0.0275	0.9	1.3	4.8	0	1.94	5
Qc24	5	0.036	0.05	5	3.5	4.8	1.3839	5	5
Qc29	5	0.03	0.024	2.4	1.42	3.3	0.0003	1.94	1.9984
Min Ploss	4.4968	4.79301	4.76601	4.6022	4.5691	4.529	4.5143	4.4984	4.4978

TABLE 6.3

Comparison Results for TVD Minimization (IEEE 30-Bus System)

Control Variables	Initial	DE (Ela, et al. 2011)	BFOA (Rajan and Malakar 2015)	PSO (Mahadevan and Kannan 2010)	FA Rajan and Malakar 2015)	HFA (Rajan and Malakar 2015)	TLBO (Ghasemi et al. 2014a)	GSA (Duman et al. 2012a)	CPVE HBMO (Ghasemi, et al 2014a)	HBMO (Ghasemi, et al. 2014)	CSA (Rajan et al. 2017)
Vg1	1.05	1.01	0.95	1.006177	0.9977	1.003458	1.0121	0.99298	1.072811	1.08276	0.9658
Vg2	1.04	0.9918	1.0702	1.007507	1.0217	1.01638	0.9806	0.95519	1.040827	1.043241	1.0395
Vg5	1.01	1.0179	0.9645	1.00839	1.01672	1.019451	1.0207	1.0189	1.037859	1.045632	1.0198
Vg8	1.01	1.0183	1.0258	1.039351	1.001	1.018221	1.0163	1.0189	1.040092	1.040956	1.0386
Vg11	1.05	1.0114	1.0375	1.003456	1.0481	0.982272	1.0293	1.01198	1.084133	1.093874	1.0494
Vg13	1.05	1.0282	0.9914	1.04466	1.0191	1.01546	1.0323	1.03598	1.022005	1.010233	1.05
T6–9	1.078	1.0265	0.98	0.99	1.04	0.99	1.0435	1.0578	0.954076	1.1	0.92
T6–10	1.069	0.9038	0.96	0.9	0.9	0.9	0.9056	1.05	1.100002	1.028374	1.05
T4–12	1.032	1.0114	1.02	0.99	0.98	0.98	1.0195	0.9	1.026029	0.987309	0.96
T27–28	1.068	0.9635	0.99	0.95	0.96	0.96	0.9492	1.05	1.000034	0	0.39
Qc10	0	4.942	4.8	4.4	3.6	3.2	0.0484	0.966	0	0	2.79
Qc12	0	1.0885	1.3	0.9	1.3	0.5	0.0066	4.5	0	4.399994	4.78
Qc15	0	4.9985	4.5	1.2	2.7	4.9	0.05	2.5	4.390634	3.332412	5
Qc17	0	0.2393	2	1.9	0.9	0.1	0.0009	1.4	3.301982	3.493845	4.96
Qc20	0	4.9958	4.3	1.1	4.2	3.8	0.05	4	3.508473	0	5
Qc21	0	4.9075	3.9	1	2.7	5	0.05	3.8	0	2.465893	5
Qc23	0	4.9863	4	0.9	3	5	0.0495	2.9	2.453422	5	4.3
Qc24	0	4.9663	4.5	1	1.7	3.9	0.0493	2.5	5	1.828498	2.72
Qc29	0	2.2325	3.4	0.9	1.7	1.5	0.0024	3.1	1.826043	1.828498	2.72
TVD		0.0911	0.149	0.1535	0.1157	0.098	0.0913	0.118	0.198756	0.210602	0.111592

(Continued)

TABLE 6.3 (*Continued*)

Comparison Results for TVD Minimization (IEEE 30-Bus System)

Control Variables	GSA-CSS (Rajan et al. 2017)	IGSA-CSS (Rajan et al. 2017)	BA (Rajan et al. 2017)	ALO (Rajan et al. 2017)	MALO (Rajan et al. 2017)	CLPSO (Duman et al. 2011)	ABC (Rajan and Malakar 2016)
Vg1	1.021714	1.008481	1.0186	1.0131	1.0049	1.1	1.0025
Vg2	1.021245	1.005722	0.9797	1.0262	0.9504	1.1	1.0161
Vg5	1.020442	1.01909	1.0193	1.0194	1.0382	1.072	0.9927
Vg8	1.007064	1.010263	1.0475	1.0264	1.0122	1.076	1.0288
Vg11	1.022921	1.018422	0.9938	0.9949	1.0406	1.045	1.0646
Vg13	1.007145	1.007997	0.9753	0.9732	1.0216	1.1	1.0086
T6–9	0.984	1.034	0.98	0.99	1.07	1.0177	0.97
T6–10	0.928	0.9	0.92	0.92	0.91	0.9738	1.03
T4–12	0.968	0.984	0.96	0.95	1.01	1.0244	0.97
T27–28	0.962	0.978	0.97	0.97	0.96	0.9896	0.95
Qc10	0.0245	0.05	3.47	4.4	3.8	0.722	2.5
Qc12	0.025	0.05	2.45	4.2	4.76	1.6812	0
Qc15	0.0365	0.05	3.37	2.6	5	2.6462	5
Qc17	0.027	0	3.63	1.1	2.26	3.4105	0
Qc20	0.044	0.05	4.34	3.7	4.84	1.9773	5
Qc21	0.0195	0.05	3.62	3.4	5	0.4767	5
Qc23	0.0355	0.05	3.41	3.6	5	3.5896	5
Qc24	0.029	0.05	4.05	3.9	5	2.9998	4.7
Qc29	0.0285	0.0495	2.35	1.9	0.58	1.1098	0
TVD	0.12394	0.08968	0.116146	0.1177	0.09707	0.245	0.135

TABLE 6.4
Comparison of Results for L-Index Minimization (IEEE 30-Bus System)

Control Variables	Initial	DE (Ela et al. 2011)	GSA (Duman et al. 2012)	OGSA (Shaw et al. 2014)	CPVEIHB-MO (Ghasemi et al. 2014a)	HBMO (Ghasemi et al. 2014a)	CSA (Rajan et al. 2017)	ALO (Rajan et al. 2017)	MALO (Rajan et al. 2017)	GWO (Sulaiman et al. 2015)	BBO (Bhattacharya and Chattopadhyay 2011)
Vg1	1.05	1.0993	1.1	1.0951	1.1	1.1	1.0999	1.1	1.1	1.0965	1.0995
Vg2	1.04	1.0967	1.1	1.0994	1.1	1.1	1.0982	1.1	1.1	1.0807	1.0822
Vg5	1.01	1.099	1.1	1.0991	1.1	1.1	1.1	1.1	1.1	1.0693	1.0738
Vg8	1.01	1.0346	1.1	1.0991	1.1	1.1	1.1	1.1	1.1	1.0624	1.0499
Vg11	1.05	1.0993	1.1	1.0995	1.1	1.1	1.1	1.1	1.1	1.0977	1.0837
Vg13	1.05	0.9517	1.1	1.0994	1.1	1.1	1.0986	1.1	1.0963	1.0927	0.96403
T6–9	1.078	0.6038	0.9	0.9728	0.9	0.9	0.92	0.919	0.92	0.96	1.0999
T6–10	1.069	0.9029	0.9	0.9	0.839363	0.9	0.93	0.905	0.91	1.01	1.0999
T4–12	1.032	0.9002	0.9	0.9534	0.895746	0.9	0.9	0.909	0.9	0.97	1.1
T27–28	1.068	0.936	1.0195	0.9501	1.023412	1.03241	0.9	0.904	0.91	0.94	0.90246
Qc10	0	0.6854	5	0.21	5	5	5	5	5	2	0.047741
Qc12	0	4.7163	5	2.65	5	5	5	5	5	1	0.049482
Qc15	0	4.4931	5	0	5	5	4.71	5	5	1	0.047491
Qc17	0	4.51	5	0.06	5	5	5	5	5	2	0.047138
Qc20	0	4.4766	5	0	5	5	4.59	4.16	5	2	0.049353
Qc21	0	4.6075	5	0	5	5	4.92	4.929	5	1	0.049498
Qc23	0	3.8806	5	0	5	5	4.87	3.343	5	4	0.049404
Qc24	0	3.8806	5	0.09	5	5	5	5	5	4	0.048298
Qc29	0	3.2741	5	0	5	5	5	5	5	4	0.048054
L-index		0.1246	0.11607	0.123	0.111029	0.11473	0.0986	0.09844	0.097851	0.118	0.09803

(Continued)

TABLE 6.4 (*Continued*)
Comparison of Results for L-Index Minimization (IEEE 30-Bus System)

Control Variables	TLBO (Ghasemi et al. 2014b)	QOTLBO (Ghasemi et al. 2014b)	ABC (Rajan and Malakar 2015)	BA (Rajan et al. 2017)	PSO (Mahadevan and Kannan 2010)	EMA (Rajan and Malakar 2016)
Vg1	1.0686	1.0956	1.0829	1.097	1.0996	1.1
Vg2	1.0747	1.0997	1.073	1.093	1.0927	1.1
Vg5	1.0696	1.0994	1.0759	1.049	1.0735	1.1
Vg8	1.08	1.0994	1.0744	1.071	1.0729	1.1
Vg11	1.0724	1.0996	1.1	1.06	0.9893	1.1
Vg13	1.095	1.1	1.0804	1.097	1.0718	1.1
T6–9	0.9385	0.9729	1.03	1.09	0.9813	0.9168
T6–10	0.9318	0.9	0.92	0.9	0.9844	0.9
T4–12	0.9498	0.9537	0.92	1.1	1.0521	0.9
T27–28	0.9331	0.9506	0.97	0.93	1.0095	0.9
Qc10	0.0395	0.0024	5	3	4.1823	5
Qc12	0.0466	0.0267	5	4	1.9829	5
Qc15	0.0392	0	5	3	3.0541	5
Qc17	0.0464	0.0005	4	5	1.2266	5
Qc20	0.0051	0	5	5	1.5015	5
Qc21	0.015	0	3	0	5	5
Qc23	0.0101	0	4	0	2.522	5
Qc24	0.0043	0.001	4	0	2.6415	5
Qc29	0.0016	0	5	3	3.2439	5
L-index	0.1252	0.1242	0.1161	0.1191	0.1089	0.09797

Figures 6.1 and 6.2 show the simulation results of different algorithms on IEEE57 bus systems used to optimize the active power loss of the ORP problem. On the other hand, Figure 6.3 represents the comparisons of different algorithms in objective function value, i.e., active power loss minimization. In this work, Rajan et al. (2017) tried to modify the recently developed ALO and applied the modified algorithms in solving ORPD problems of the power system. The authors have critically observed the convergence phenomena of ALO and found that the algorithm gets trapped into local optima. To improve the convergence characteristics of the algorithm, it is essential to have the right balance between its exploitation and exploration phase. The authors have introduced the weight factor in the original algorithm to balance both, which can be optimally tuned for different applications, and better results can be obtained. The results of modified algorithms are also compared with other population-based algorithms like ALO, CSA, BA, ABC, PSO, and FA. The convergence plots of all three cases are also presented in Figure 6.4. From the figures, it can be seen that modified ALO (MALO) converges fast and is able to beat several other algorithms

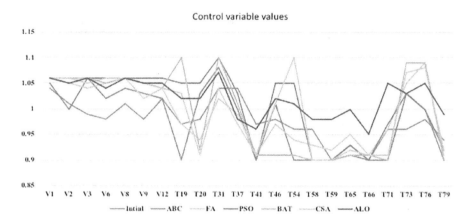

FIGURE 6.1 Optimal settings of generator voltage and transformers for P loss minimization (IEEE 57-bus system).

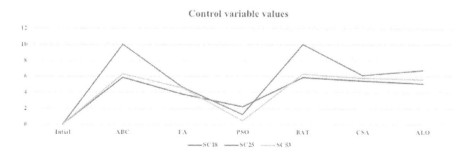

FIGURE 6.2 Optimal settings of shunt capacitors for P loss minimization (IEEE 57 bus system).

Optimum Scheduling with Renewable Integration

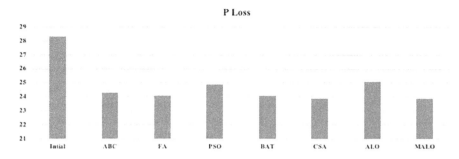

FIGURE 6.3 Optimal values of P loss with different algorithms (IEEE 57-bus system).

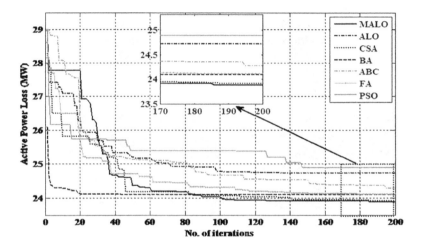

FIGURE 6.4 Convergence plots of different algorithms for IEEE 57-bus system.

as far as the optimization of objective functions is concerned. For more details, the article by Rajan et al. (2017) can be studied.

6.5 OPTIMAL ACTIVE POWER DISPATCH

In recent times, the economy has begun to play an active role for the policymakers on how a power system has to be planned and operated. In view of this, the power system has been a potent application of suitable computational methods or algorithms. The algorithm/computational approach aims to determine a practical or feasible solution for power system operation quickly. Until today, major sources of electricity are thermal power generating stations. Because of the faster decline in conventional energy sources such as coal, emphasis is now being given to running the coal-based generation most economically. Researchers in the past had given stresses only on economic aspects while scheduling the generators. Still, recent environmental regulations imposed by the world body limit emissions caused by thermal generation. Therefore, along with cost, restricting emissions from thermal generators also gained premium importance.

The power system operator has to be cost-effective to ensure a cost-efficient electrical system. The primary requirement of such a system is to serve the consumers' demand in real-time satisfactorily by adjusting the power output of generating units most economically. This exercise is termed the optimal active power dispatch (OAPD) problem in the power system terminology. It is an optimization problem (Wood and Wollenberg 1996) that ensures the economic dispatch of active power from all the committed units by meeting all the constraints related to the generation and transmission of real power.

The control variables in OAPD problems are power outputs of thermal generators. These variables are controlled through optimization techniques under all the constraints mentioned above to minimize the power generation cost. The OAPD problems can be solved as an OPF problem. Scheduling is a process of allocation of generations among different generating units. When the scheduling is performed using OPF, it is considered a power flow constrained GS problem. The most commonly used objective functions for OAPD problems are the minimization of fuel cost for thermal units (Lee et al. 1985, Granville 1994, Goldberg 1989, 2002) and the minimization of environmental emissions.

6.5.1 Problem Formulation of OAPD problems

Dependent variable: slack bus active (P_{G1}), and reactive power (Q_{G1}), load voltages $\left(V_{L1}, \ldots V_{L_{NPQ}}\right)$, reactive power generations $\left(Q_{G1}, \ldots Q_{G_{NPV}}\right)$, and line total power $\left(S_{L1}, \ldots S_{L_{NTL}}\right)$.

The vector of dependent variables 'x':

$$x^T = \left[P_{G1}, V_{L1}, \ldots V_{L_{NPQ}}, Q_{G1}, \ldots Q_{G_{NPV}}, S_{L1}, \ldots S_{L_{NTL}}\right] \tag{6.19}$$

The vector of independent/ control variables 'u'

$$u^T = \left[\overbrace{P_{G2}, \ldots P_{G_{NPV}}, V_{G1}, \ldots V_{G_{NPV}}}^{\text{continuous}} \overbrace{\text{Tap}_1, \ldots \text{Tap}_{NT}, Q_{C_1}, \ldots Q_{C_{NC}}}^{\text{discrete}}\right] \tag{6.20}$$

The objective considered in OAPD problems are as follows:

1. Fuel cost minimization
   a. Simple fuel cost
      Simple fuel cost can mathematically be represented as follows:

$$F_{\text{cost}}^{\text{simple}} = \sum_{i=1}^{i=N_g} \left[c_i + b_i P_{Gi} + a_i P_{Gi}^2 + \right] \text{\$/h}. \tag{6.21}$$

The terms a_i, b_i and c_i are cost coefficients of the ith thermal generator, and N_g is the total generators present. This is the simplest form of cost curve. The graphical presentation of (6.21) is shown in Figure 6.5. From this figure, it can be seen that the nature of simple fuel cost is convex and continuous.

Optimum Scheduling with Renewable Integration

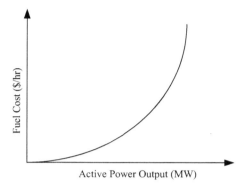

FIGURE 6.5 Graphical representation of simple fuel cost.

b. Fuel cost minimization with valve point loading effect:
   Mathematically, the cost function with valve point loading (VPL) is expressed as (Niknam et al. 2012)

$$F_{cost}^{valve\ point} = \sum_{i=1}^{i=N_g}\left[c_i + b_i P_{Gi} + a_i P_{Gi}^2 + \left|e_i \sin\left(f_i\left(P_{Gi}^{min} - P_{Gi}\right)\right)\right|\right] \$/h. \quad (6.22)$$

e_i and f_i are VPL coefficients. VPL effect can be seen from Figure 6.6.

2. Environmental emission optimization.

Pollutants in the environment consists of nitrogen oxides (NO_x), sulfur oxides (SO_x) and carbon oxides (CO_x), which are emitted in large amounts during the burning of fuels. Hence, apart from economy, environmental emission minimization is equally important. In this chapter, the expression used for minimization of environmental emission is presented below (Basu 2011):

$$F_{emission}^{total} = 10^{-2} \times \left(\alpha_i + \beta_i P_i + \gamma_i P_i^2\right) + \zeta_i \exp\left(\lambda_i P_i\right) \text{ton/h} \quad (6.23)$$

where α_i, β_i, γ_i, ζ_i and λ_i are coefficients generator emission characteristics.

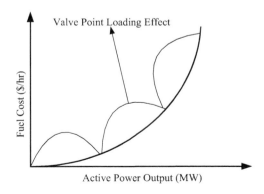

FIGURE 6.6 Graphical representation of VLP effect.

- Related work

  Many traditional techniques (Lee et al. 1985, Zehar and Sayah 2008, Alsac et al. 1990) were implemented in solving OAPD problems. Lee et al. (1985) proposed unified gradient projection method for real and reactive power dispatch. Zehar and Sayah (2008) solved an environmental/economic load dispatch problem on a multiobjective framework using successive linear programming (LP) technique. The Algerian 59-bus power system was considered as test environment. The LP technique is modeled as OPF and discussed by Alsac et al. (1990). In this work, the authors discussed the application of the LP-based method in modeling nonseparable objective functions. These classical optimization techniques have the advantage of a fast convergence rate. If the function is unimodal or linear, these algorithms provide the optimal solution to the problem. On the contrary, when the objective function becomes nonlinear or multimodel, they never guarantee the optimal solution. Hence, researchers have started diverting their interest to general-purpose optimization techniques as substitutes in the past few decades. In terms of optimization, these algorithms are termed meta-heuristic algorithms. Some examples of meta-heuristic algorithms are GA, ACO, PSO, DE, etc. These algorithms are general-purpose algorithms, and they work independently to the nature of the objective function.

  Algorithms like GA (Devaraj and Yegnanarayana 2005, Iba 1994, Lai et al. 1997, Paranjothi and Anburaja 2002, Bakirtzis et al. 2002, Attia et al. 2012), Evolutionary Programming (EP) (Reddy et al. 2014, Yuryevich and Wong 1999), Tabu Search (TS) (Abido 2002), PSO (Vaisakh et al. 2013), DE (Ela et al. 2010), BBO (Bhattacharya and Chattopadhyay 2011), HSA (Sivasubramani and Swarup 2011), BHBA (Bouchekara et al. 2014), TLBO (Krishnanand et al. 2013), etc. have performed well in solving many electrical engineering optimization problems. Lai et al. (1997) developed GA with the dynamical hierarchy of the coding system, and the OPF problem is solved. Enhanced GA (EGA) (Bakirtzis et al. 2002) solves OPF for the first time with continuous and discrete decision variables. Abido solved the OPF problem using TS algorithm. PSO (Vaisakh et al. 2013) gained much popularity in solving complex optimization problems because the algorithm is simple to understand and easy to implement. Many researchers have tried PSO in solving different OPF problems. M. A. Abido (2014) utilized PSO in solving the OPF problem of the power system. Both fuel cost minimization and voltage security-related objectives are considered. Problem is implemented on the IEEE 30-bus system, and results are compared with other reported literature. Bhattacharya and Chattopadhyay utilized BBO techniques to solve the OPF problem (Bhattacharya and Chattopadhyay 2011), whereas, Sivasubramani and Swarup (2011) solved generation scheduling problem with the aim to optimize both fuel cost and environmental emission using HAS. Ghasemi et al. (2014b) solved the OPF problems of power systems.

Optimum Scheduling with Renewable Integration

The results are compared with its other variants and several other methods reported to solve the same problem.

After critically reviewing the application of this meta-heuristic algorithm, researchers found that these algorithms also tend to trap into local optima and slower convergence. To improve these algorithms' convergence and search capability, researchers have tried to hybridize them either with classical optimization techniques or with some intelligent approaches. These hybrid algorithms also find their application in solving power system optimization problems (Niknam et al. 2012). Ghasemi et al. (2014a) proposed SFLA, and Simulated Annealing (SA) was proposed by Niknam et al. for solving nonsmooth OAPD problem. The generator operational constraints such as VPL and prohibited operating zone (POZ) are also considered in this work. Rajan and Malakar (2015) combined FA with Nelder-Mead (NM) local search technique to solve the OAPD problem of the power system. Both VPL and POZ are considered in this work. The results depict that the convergence curve of FA is improved with the addition of NM as a local search subroutine. Similarly, Panigrahi and Pandi (2008) combined NM with BFO to obtain better convergence. The generation scheduling problem is solved, and results are compared with the original version of BFOA and several other reported methods. Varadarajan and Swarup (2008) considered environmental emission as one of the objectives and fuel cost, and the problem is modeled as an OPF problem. DE is used as an optimization tool.

- Comparative results for fuel cost minimization

  Table 6.5 shows the results obtained from different algorithms in minimizing the fuel cost of IEEE 30-bus system. All system information are taken from Niknam et al. (2012) (Table 6.6).

Results infer that the meta-heuristic algorithms are giving probable solutions the complex optimization problems which cannot be solved by the deterministic method.

6.6 OPTIMAL REACTIVE POWER DISPATCH WITH RENEWABLE POWER GENERATIONS

In the present-day scenario, since thermal-based power plants are limiting its energy production due to day by day depleting high-quality coal in India and other leading coal producing countries, penetration of renewable energy sources (RESs) into the grid is gaining attention day by day. The significant advantage of renewable energy source is that it's free of cost availability, but it is very intermittent. Apart from active power dispatch problems, reactive power dispatch problems are also solved in the presence of renewable power generations (RPGs). Reactive power control becomes a concerning issue with the extensive penetration of RPGs. As most RPGs take reactive power prom grid, large penetration of such power plants may lead to voltage instability or, in the worst case, voltage collapse. Among various RPGs, wind and solar are gaining popularity among researchers. The classification and mathematical modeling of both the power generating systems are presented below.

TABLE 6.5

Comparative Results of Simple Fuel Cost Minimization for IEEE 30-Bus System

GM (Lee et al. 1985)	MDE (Mahadevan and Kannan, 2010)	IGA (Lai et al. 1997)	PSO (Abido 2014)	DE	BBO (Bhattacharya and Chattopadhyay 2011)	EEA (Arora and Singh 2019)	MOHSA (Sivasubramani and Swarup 2011)	De-OPF-RGA (Paranjothi and Anburaja 2002)
804.853	802.306	800.805	800.41	799.289	799.1116	800.0831	798.8	806.86

GA (Paranjothi and Anburaja 2002)	EP (Yuryevich and Wong 1999)	EGA (Bakirtzis et al. 2002)	MICA-TLA (Ghasemi et al. 2014c)	TS (Abido 2002)	BHBA (Bouchekara 2014)			
805.94	802.62	802.06	801.0488	802.29	799.9217			

TABLE 6.6

Comparative Results of Fuel Cost Minimization with VPL for IEEE 30-Bus System

GA (Niknam et al. 2012)	PSO (Niknam et al. 2012)	SA (Niknam et al. 2012)	SFLA (Niknam et al. 2012)	HSFLA-SA (Niknam et al. 2012)	Hybrid GA1 (Mezura-Montes and Coello 2011)	Hybrid GA2 (Reddy and Bijwe 2016)	Combined GA (Reddy and Bijwe 2016)	EGA (Reddy et al. 2014)	Proposed EEA (Reddy et al. 2014)
829.44931	826.5897702	827.82629	825.99061	825.6922	826.8492	826.7554	826.6962	826.3176	826.8492

6.6.1 CLASSIFICATION OF WIND TURBINES

The widely used technology used for the common application of wind turbines are horizontal-axis wind turbines. These wind turbines are generally nondispatchable unit as it is difficult to follow the load due to the fluctuating nature of wind speed and hence the power. Generally, the power output of these type of turbines ranges from 500 kW to 5 MW. A wind turbine for power generation consists of following sub systems:

- Rotor circuit of wind turbines
- Electrical generator, gearbox, shafts, couplings, and mechanical brake
- Tower
- Foundation
- Cables, transformers, and power electronic devices.

The complete block diagram is provided in Figure 6.7 (Singh and Santoso 2011). A wide variety of horizontal-axis wind turbines is used today. Different varieties of wind turbines have different establishment costs, complexity, and efficiency of power generation. As mentioned above, they consist of blade and rotor structures to extract wind power; gear trains are used to vary the speed input to the generator and induction generator for electromechanical energy conversion. Induction generators are asynchronous machines and are very useful for power generation with variable speed. Power electronics devices regulate the need for reactive power to induction generators and active power to the greed.

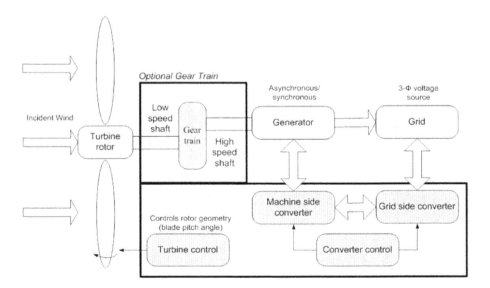

FIGURE 6.7 Complete wind turbine model.

Wind turbines are classified into four basic types:

a. Fixed speed wind turbines: It the basic wind turbine used for wind power generations. Squirrel cage induction generators are used in electromechanical energy conversion. They are relatively robust and reliable. The major disadvantages of this type of turbine are their low efficiency, limit range of speed variation, and requirement of reactive power compensation.
b. Variable speed wind turbines: It is designed to operate with wide range of rotor speed. They have both speed and power control which helps in extracting more energy from wind. Rotor resistance control are available which allows wide range of slip variation. However, some power is also lost in the rotor resistance.
c. Doubly fed induction generator (DFIG) wind turbines: DFIGs are designed to overcome the drawbacks of variable speed wind turbines. They employ back-to-back AC/DC/AC convertors in the rotor as slip power recovery mechanism. The wind power extraction is high with less mechanical stress which increases the efficiency of the machine. Rotor currents allows decoupled real and reactive power outputs due to flux-vector control.
d. Full-converter wind turbines: Full-converter wind turbines use back-to-back AC/DC/AC convertors which is the only path for power to flow from generation to the grid. Hence, due to no direct connection from wind turbine to grid it also allows synchronous generators as the electromechanical energy conversion device for active and reactive power control.

The different wind turbine technologies discussed above are shown in Figure 6.8 (Singh and Santoso 2011).

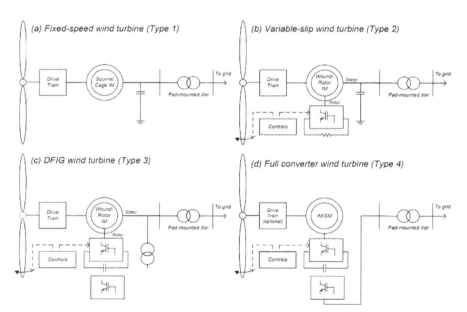

FIGURE 6.8 Different technologies used for wind turbines.

Optimum Scheduling with Renewable Integration

1. *Modeling of wind power generations*

   Wind Speed distribution generally follow the Weibull probability distribution function (Zhihuan et al. 2010). The probability density of the wind speed (v_w) is given as

   $$\Delta_v(v_w) = \left(\frac{\beta}{\alpha}\right)\left(\frac{v_w}{\alpha}\right)^{(\beta-1)} \exp\left[-\left(\frac{v_w}{\alpha}\right)^{\beta}\right] \text{ for } 0 < v_w < \infty \qquad (6.24)$$

   α: scale parameter
   β: is the shape parameter for Weibull PDF.

2. *Modeling of solar power generations*

   PV system is a promising option among the available RESs. The produced electrical output from the PV system is dependent on environmental conditions like, solar irradiance, ambient temperature etc. Additionally, as the PV system is semiconductor based, the system is nonlinear in nature. Partial shading issue adds more complicacy in operation. The single-diode modeling of PV system is used which can be elaborated as follows (Figure 6.9):

I_{PV} = Current source of PV system.
D = Diode connected in parallel with the current source.
R_{se} = Sum of resistances of the components in series path of current.
R_p = This resistance represents the leakage across the PV cell.
I_D = Diode current.
I = Difference between photo current (I_{PV}) and I_D.

$$I = I_{PV} - I_0\left[\exp\left(\frac{qV + qR_{se}I}{N_s k_s Ta} - 1\right)\right] - \frac{V + R_{se}I}{R_p} \qquad (6.25)$$

I_0 = Saturation current,
q = Charge of electron,
T = Temperature in kelvin,
a = Diode ideality factor,
I_D = Boltzmann's constant,
N_s = Number of cells in series.

The lognormal probability density with mean (μ_s) and standard deviation (σ_s) of solar irradiance (G_s) is represented as

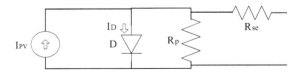

FIGURE 6.9 Electrical circuit of single-cell PV panel.

$$\Delta_v(v_w) = \frac{1}{G_s \sigma_s \sqrt{2\pi}} \exp\left[\frac{-(\ln G_s - \mu_s)^2}{2\sigma_s^2}\right] \tag{6.26}$$

Several other probabilistic models for PV generation are discussed by Prusty and Jena (2016, 2018).

In view of the above discussion, many researchers have solved the ORPD problems of power system considering the uncertain nature of RPGs. Some of them are summarized below:

References	Contributions
Hu et al. (2010)	Power injections at node and branch random outages are considered as the uncertainty sources.
Zhihuan et al. (2010)	A stochastic MO-ORPD (SMO-ORPD) problem is solved in a wind integrated system with loads and wind power generation as uncertainties.
Mohseni-Bonab et al. (2015, 2016a)	Active power loss and total voltage deviation are considered as the objective functions in MO-ORPD, and treatment of uncertain nature of load is done by Monte Carlo simulation and two-point estimate method.
Shargh et al. (2016)	Probabilistic multiobjective OPF considering correlation in wind speed and load is presented.
Liang et al. (2015)	Enhanced firefly algorithm developed for multiobjective optimal active and reactive power dispatch problem. Load and wind power generation uncertainties were taken into consideration.
Mohseni-Bonab et al. (2016b)	Stochastic multiobjective ORPD (SMO-ORPD) was analyzed with the aim of minimizing voltage stability and real power loss. Wind power generation uncertainties were integrated into the SMO-ORPD while solving it.
Mohseni-Bonab et al. (2017)	Two objectives were considered: minimization of real power loss and operation and maintenance cost of windfarm for ORPD problem. The ORPD problem considered voltage deviation, real power loss, and voltage stability index as the objectives of optimization.

6.7 ACTIVE AND REACTIVE POWER SHARING IN MICROGRID

Microgrids consist of various distributed generators (DGs) connected in parallel with highly sophisticated control mechanisms that can operate both with and without a grid (islanded mode). Microgrid helps in alleviating the stress of the main transmission line, losses in the feeders. Moreover, it is highly helpful in maintaining the quality of active and reactive power. As far as the islanded modes are concerned, it is essential to achieve the system's stability by effective sharing of loads among the connected DGs. Since different DGs offer different internal impedances while connected in parallel, sharing active and reactive power loads is not proportional to their capacity due to their internal impedance mismatch. Moreover, they are small units with different power ratings, which also adversely influence the sharing of active and reactive power loads among the DGs. The reason behind this poor sharing of powers among DGs is the traditional droop control schemes. Therefore, some advanced

Optimum Scheduling with Renewable Integration

controlling mechanisms have been adopted to overcome the issues of unequal sharing of loads. These mechanisms are

1. Adaptive/improved droop control
2. Network-based control methods
3. Cost-based droop schemes.

Furthermore, due to the nonlinear and unbalanced load, the reactive power is adversely affected while controlling the active power. Hence, it is challenging to share the reactive power data in a correct manner with the conventional virtual impedance method. As such, the hierarchical control approaches are exploited as supplements of the orthodox droop controls and virtual impedance methods. The enhanced hierarchical control approaches like the algorithms based on predictive control, graph theory, multiagent system, and gain scheduling method are some improved methods for the equal reactive power load sharing mechanism.

6.8 CONCLUSION

This chapter deals with the different methods and solution algorithms used to solve the OPF problem of the power system. OPF is the power system optimization tool that tries to optimize specific objective functions simultaneously, fulfilling all the operations and physical constraints of the power system. This chapter discusses the importance of solving two major optimization problems of power systems, i.e., optimal active and optimal reactive power dispatch problems. Since these are highly nonlinear nonconvex optimization problems, multipurpose robust optimization techniques are used as the solution techniques. The chapter also attempts to discuss OPF problems with different meta-heuristic algorithms to solve both active and reactive power dispatch problems. Results compared with many contemporary algorithms. In the past few decades, RPGs are gaining importance in dispatching loads during peak hours. In view of this, the current chapter discusses the methodology used to integrate RPGs such as wind and solar power generation to the grid. Difficulties faced by the researchers in modeling the intermittent nature of these RPGs are broadly described in this chapter. Some improved methods for proper active and reactive power loads sharing of RPGs in islanding mode are also highlighted in the present chapter.

REFERENCES

The IEEE 30-Bus Test System. http://www.ee.washington.edu/research/pstca/pf30/pg_ca30bus. htm. (Accessed 23rd March, 2021).

The IEEE 118-Bus Test System. http://www.ee.washington.edu/research/pstca/pf118/pg_tca118bus.htm. (Accessed 23rd March, 2021).

Abido, M.A. 2002. Optimal power flow using tabu search algorithm. *Electric Power Components and Systems.* 30(5):469–483.

Abido, M. A. 2014. Optimal power flow using particle swarm optimization. *International Journal of Electrical Power and Energy Systems.* 24(7):563–571.

Alsac, O., J. Bright, M. Prais, and B. Stott. 1990. Further development in LP- based optimal power flow. *IEEE Transactions on Power Systems.* 5(3):697–711.

Arora, S., and S. Singh. 2019. Butterfly optimization algorithm: A novel approach for global optimization. *Soft Computing*. 23:715–734.

Attia, A-F, Y. A. Al-Turki, and A. M. Abusorrah. 2012. Optimal power flow using adapted genetic algorithm with adjusting population size. *Electric Power Components and Systems*. 40(11):1285–1299.

Bakirtzis, A. G., P.N. Biskas, C.E. Zoumas, and V. Petridis. 2002. Optimal power flow by enhanced genetic algorithm. *IEEE Transactions on Power Systems*. 17(2):229–236.

Basu, M. 2011. Economic environmental dispatch using multi-objective differential evolution. *Applied Soft Computing*. 11(2): 2845–2853.

Bhattacharya, A., and P. K. Chattopadhyay. 2011. Application of biogeography-based optimisation to solve different optimal power flow problems. *IET Generation, Transmission & Distribution*. 5(1):70–80.

Bhattacharya, A., and P. K. Chattopadhyay. 2010. Solution of optimal reactive power flow using biogeography-based optimization. *International Journal of Electrical and Computer Engineering*. 4(3):621–629.

Biswas, S., T. Som, K. K. Mandal and N. Chakraborty. 2014. Cuckoo search algorithm based optimal reactive power dispatch. In *Proceedings of the International Conference on Control, Instrumentation, Energy & Communication*, Calcutta, India; 2014, 412–416.

Bjelogrlic, M., M. S. Calovic, P. Ristanovic and B.S. Babic. 1990. Application of Newton's optimal power flow in voltage/reactive power control. *IEEE Transactions on Power Systems*. 5(4):1447–1454.

Bouchekara, H. R. E. H. 2014. Optimal power flow using black-hole-based optimization approach. *Applied Soft Computing*. 24:879–888.

Carpentier, J. 1979. Optimal power flows. *International Journal of Electrical Power and Energy Systems*. 1 (1):3–15.

Dai, C., W. Chen, Y. Zhu and X. Zhang. 2009. Seeker optimization algorithm for optimal reactive power dispatch. *IEEE Transactions on Power Systems*. 24(3):1218–1231.

Devaraj, D., and B. Yegnanarayana. 2005. Genetic algorithm based optimal power flow for security enhancement. *IEE Proceedings - Generation, Transmission and Distribution*. 152(6): 899–905.

Dommel, H. W., and W. F. Tinney. 1968. Optimal power flow solutions. *IEEE Transactions on Power Apparatus and Systems*. PAS-87(10):1866–76.

Dorigo, M. 1992. *Optimization, learning and natural algorithms*. Ph.D. thesis: Politecnico di Milano, Italy.

Duman, S., U. Güvenç, Y. Sönmez, and N. Yörükeren. 2012a. Optimal power flow using gravitational search algorithm. *Energy Conversion and Management*. 59:86–95.

Duman, S., Y. Sonmez, U. Guvenc and N. Yorukeren. 2011. Application of gravitational search algorithm for optimal reactive power dispatch problem. *International Symposium on Innovations in Intelligent Systems and Applications,* Istanbul, Turkey, 519–523.

Duman, S., Y. Sonmez, U. Guvenc and N. Yorukeren. 2012b. Optimal reactive power dispatch using a gravitational search algorithm. *IET Generation, Transmission and Distribution*. 6(6):563–576.

Ela, A. A. A. E., M. A. Abido, and S. R. Spea. 2010. Optimal power flow using differential evolution algorithm. *Electric Power Systems Research*. 80(7):878–885.

Ela, A. A. A. E., M. A. Abido, S. R. Spea. 2011. Differential evolution algorithm for optimal reactive power dispatch. *Electric Power Systems Research*. 81(2):458–464.

Eskandar, H., A. Sadollah, A. Bahreininejad, and M. Hamdi. 2012. Water cycle algorithm– A novel metaheuristic optimization method for solving constrained engineering optimization problems. *Computers & Structures*. 110–111:151–166.

Fatah, S. A., M. Ebeed and S. Kamel. 2019. Optimal reactive power dispatch using modified sine cosine algorithm. *2019 International Conference on Innovative Trends in Computer Engineering*, Aswan, Egypt, 2019.

Ghasemi, A., K. Valipour and A. Tohidi. 2014a. Multi objective optimal reactive power dispatch using a new multi objective strategy. *International Journal of Electrical Power and Energy Systems*. 57:318–334.

Ghasemi, M., M. M. Ghanbarian, S. Ghavidel, E. M. Moghaddam. 2014b. Modified teaching learning algorithm and double differential evolution algorithm for optimal reactive power dispatch problem: A comparative study. *Information Sciences*. 278:231–149.

Ghasemi, M., S. Ghavidel, S. Rahmani, A. Roosta, and H. Falah, 2014c. A novel hybrid algorithm of imperialist competitive algorithm and teaching learning algorithm for optimal power flow problem with non smooth cost functions. *Engineering Applications of Artificial Intelligence*. 29:54–69.

Goldberg, D.E. 1989. *Genetic Algorithms in Search, Optimization, and Machine Learning*. Addison-Wesley Longman Publishing Co, Boston, MA.

Goldberg, D.E. 2002. *The Design of Innovation: Lessons from and for Competent Genetic Algorithms*. Kluwer Academic Publishers, Dordrecht.

Granville, S. 1994. Optimal reactive dispatch through interior point methods. *IEEE Transactions on Power Systems*. 9(1):136–146.

Han, Y., H. Li, P. Shen, E. A. A. Coelho, and J. M. Guerrero. 2017. Review of active and reactive power sharing strategies in hierarchical controlled microgrids. *IEEE Transactions on Power Electronics*. 32(3):2427–2451.

Hu, Z., X. Wang, and G. Taylor. 2010. Stochastic optimal reactive power dispatch: Formulation and solution method. *International Journal of Electrical Power and Energy Systems*. 32(6):615–621.

Huang, C.M., S.J. Chen, Y.C. Huang and H.T. Yang. 2012. Comparative study of evolutionary computation methods for active–reactive power dispatch. *IET Generation, Transmission and Distribution*, 6(7):636–645.

Iba, K. 1994. Reactive power optimization by genetic algorithm. *IEEE Transactions on Power Systems*. 9(2):685–692.

Jangir, P., S. A. Parmar, I. N. Trivedi and R. H. Bhesdadiya. 2016. A novel hybrid particle swarm optimizer with multi verse optimizer for global numerical optimization and optimal reactive power dispatch problem. *Engineering Science and Technology, an International Journal*. 20(2):570–586.

Kennedy, J., and R. Eberhart. 1995. Particle swarm optimization. In *Proceedings of the IEEE International Conference on Neural Networks*, Piscataway, NJ, USA; 1995, 1942–1948.

Khorsandi, A., A. Alimardani, B. Vahidi, S. H. Hosseinian. 2010. Hybrid shuffled frog leaping algorithm and nelder–mead simplex search for optimal reactive power dispatch. *IET Generation, Transmission and Distribution*. 5(2):249–256.

Kinnear Jr., K.E. 1994. A perspective on the work in this book. In *Advances in Genetic Programming*, edited by K. E. Kinnear Jr., 3–19. MIT Press, Cambridge, MA.

Krishnanand, K. R., S. M. F. Hasani, B. K. Panigrahi and S. K. Panda. 2013. Optimal power flow solution using self–evolving brain–storming inclusive teaching–learning–based algorithm. In *Advances in Swarm Intelligence*, ICSI 2013, Lecture Notes in Computer Science, vol. 7928, edited by Y. Tan, Y. Shi, H. Mo, 338–345. Springer-Verlag, Berlin.

Lai, L. L., J. T. Ma, R. Yokoyama, and M. Zhao. 1997. Improved genetic algorithms for optimal power flow under both normal and contingent operation states. *International Journal of Electrical Power and Energy Systems*. 19(5):287–292.

Lee, K.Y., Y. M. Park, J. L. Ortiz. 1985. A united approach to optimal real and reactive power dispatch. *IEEE Transactions on Power Apparatus and Systems*. 104(5):1147–1153.

Liang, R-H., J.-C. Wang, Y.-T. Chen, and W.-T. Tseng. 2015. An enhanced firefly algorithm to multi-objective optimal active/reactive power dispatch with uncertainties consideration. *International Journal of Electrical Power and Energy Systems*. 64:1088–1097.

Mahadevan, K. and P.S. Kannan. 2010. Comprehensive learning particle swarm optimization for reactive power dispatch. *Applied Soft Computing*. 10(2):641–652.

Mezura-Montes, E., and C. A. C. Coello. 2011. Constraint-handling in nature inspired numerical optimization: past, present and future. *Swarm and Evolutionary Computation.* 1(4):173–194.

Mohseni-Bonab, S.M., A. Rabiee, and B. Mohammadi-Ivatloo. 2016. Voltage stability constrained multiobjective optimal reactive power dispatch under load and wind power uncertainties: A stochastic approach. *Renewable Energy.* 85:598–609.

Mohseni-Bonab, S.M., A. Rabiee, and B. Mohammadi-Ivatloo. 2017. Multi-objective optimal reactive power dispatch considering uncertainties in the wind integrated power systems. In *Reactive Power Control in AC Power Systems,* edited by N. M. Tabatabaei, A. J. Aghbolaghi, N. Bizon, F. Blaabjerg, 475–513. Springer, Berlin.

Mohseni-Bonab, S.M., A. Rabiee, S. Jalilzadeh, B. Mohammadi-Ivatloo, and S. Nojavan. 2015. Probabilistic multi objective optimal reactive power dispatch considering load uncertainties using monte carlo simulations. *Journal of Operation and Automation in Power Engineering.* 3(1):83–93.

Mohseni-Bonab, S.M., A. Rabiee, B. Mohammadi-Ivatloo, S. Jalilzadeh, and S. Nojavan. 2016. A two-point estimate method for uncertainty modeling in multi-objective optimal reactive power dispatch problem. *International Journal of Electrical Power and Energy Systems.* 75: 194–204.

Niknam, T., M. R. Narimani, and R. Azizipanah-Abarghooee. 2012. A new hybrid algorithm for optimal power flow considering prohibited zones and valve point effect. *Energy Conversion and Management.* 58:197–206.

Panigrahi, B. K, and V. R. Pandi. 2008. Bacterial foraging optimization: Nelder–mead hybrid algorithm for economic load dispatch. *IET Generation, Transmission and Distribution.* 2(4):556–565.

Paranjothi, S. R. and K. Anburaja. 2002. Optimal power flow using refined genetic algorithm. *Electrical Power Components and Systems.* 30(10):1055–1063.

Prusty, B. R., and D. Jena. 2016. Combined cumulant and Gaussian mixture approximation for correlated probabilistic load flow studies: a new approach. *CSEE Journal of Power and Energy Systems.* 2(2): 71–78.

Prusty, B. R., and D. Jena. 2018. Preprocessing of multi-time instant PV generation data. *IEEE Transactions on Power System*s. 33(3):3189–3191.

Rajan, A., K. Jeevan and T. Malakar. 2017. Weighted elitism-based ant lion optimizer to solve optimum var planning problem. *Applied Soft Computing.* 55:352–370.

Rajan, A., and T. Malakar. 2016. Exchange market algorithm based optimum reactive power dispatch. *Applied Soft Computing.* 43:320–336.

Rajan, A, and T. Malakar. 2015. Optimal reactive power dispatch using hybrid nelder–mead simplex based firefly algorithm. *International Journal of Electrical Power and Energy Systems.* 66:9–24.

Reddy, S. S., and P.R. Bijwe. 2016. Efficiency improvements in meta-heuristic algorithms to solve the optimal power flow problem. *International Journal of Electrical Power and Energy Systems.* 82:288–302.

Reddy, S. S., P. R. Bijwe, and A. R. Abhyankar. 2014. Faster evolutionary algorithm based optimal power flow using incremental variables. *International Journal of Electrical Power and Energy Systems.* 54:198–210.

Runarsson, T. P., and X. Yao. 2000. Stochastic ranking for constrained evolutionary optimization. *IEEE Transactions on Evolutionary Computation.* 4(3):284–294.

Sayah, S. and K. Zehar. 2008. Modified differential evolution algorithm for optimal powerflow with non-smooth cost functions. *Energy Conversion and Management.* 49(11):3036–3042.

Schoenauer, M. and Z. Michalewicz. 1996. Evolutionary computation at the edge of feasibility. In *Parallel Problem Solving from Nature – PPSN IV,* edited by H.M. Voigt., W. Ebeling, I. Rechenberg, H. P. Schwefel. 245–254. Springer, Berlin.

Optimum Scheduling with Renewable Integration

Shargh, S., B. Mohammadi-Ivatloo, H. Seyedi, and M. Abapour. 2016. Probabilistic multi-objective optimal power flow considering correlated wind power and load uncertainties. *Renewable Energy*. 94:10–21.

Shaw, B., V. Mukherjee, S. P. Ghoshal. 2014. Solution of reactive power dispatch of power systems by an opposition-based gravitational search algorithm. *International Journal of Electrical Power and Energy Systems*. 55:29–40.

Singh, M. and S. Santoso. 2011. *Dynamic Models for Wind Turbines and Wind Power Plants- A project report*. The University of Texas at Austin, Texas.

Sivasubramani, S., and K. S. Swarup. 2011. Multi-objective harmony search algorithm for optimal power flow problem. *Electrical Power and Energy System*. 33(3):745–752.

Storn, R., and K. Price. 1997. Differential evolution: A simple and efficient heuristic for global optimization over continuous spaces. *Journal of Global Optimization*. 11:341–359.

Subbaraj, P., and P.N. Rajnarayanan. 2009. Optimal reactive power dispatch using self-adaptive real coded genetic algorithm. *Electric Power Systems Research*. 79(2):374–381.

Sulaiman, M. H., Z. Mustaffa, M. R. Mohamed and O. Aliman. 2015. Using the gray wolf optimizer for solving optimal reactive power dispatch problem. *Applied Soft Computing*. 32:286–292.

Ullah, A. S. S. M. B., R. Sarker, and C. Lokan.2012. Handling equality constraints in evolutionary optimization. *European Journal of Operational Research*. 221(3):480–490.

Vaisakh, K., L. R. Srinivas, and K. Meah. 2013. Genetic evolving ant direction particle swarm optimization algorithm for optimal power flow with non-smooth cost functions and statistical analysis. *Applied Soft Computing*. 13(12):4579–4593.

Varadarajan, M., and K. S. Swarup. 2008. Solving multi-objective optimal power flow using differential evolution. *IET Generation, Transmission and Distribution*. 2(5):720–730.

Wolpert, D. H., and W. G. Macready. 1997. No free lunch theorems for optimization. *IEEE Transactions on Evolutionary Computation*. 1(1):67–82.

Wood, A.J., and B.F. Wollenberg. 1996. *Power Generation, Operation and Control*. John Wiley and Sons Inc., Hoboken, NJ.

Wu, G., W. Pedrycz, P. N. Suganthan, and R. Mallipeddi. 2015. A variable reduction strategy for evolutionary algorithms handling equality constraints. *Applied Soft Computing*. 37:774–786.

Yuryevich, J. and K. P. Wong. 1999. Evolutionary programming based optimal power flow algorithm. *IEEE Transactions on Power Systems*. 14(4):1245–1250.

Zehar, K. and S. Sayah. 2008. Optimal power flow with environmental constraint using a fast successive linear programming algorithm- application to the Algerian power system. *Energy Conversion and Management*. 49(11):3361–3365.

Zhihuan, L., L. Yinhong, and D. Xianzhong. 2010. Non-dominated sorting genetic algorithm-II for robust multi-objective optimal reactive power dispatch. *IET Generation, Transmission and Distribution*. 4(9):1000–1008.

7 Role of Stochastic Optimization for Power System Operation and Decision-Making

Wen-Shan Tan
Monash University Malaysia

Mohamed Abdel Moneim Shaaban
Universiti Malaya

CONTENTS

Nomenclature 165
7.1 Stochastic Optimization 166
7.2 Power System Operation with Stochastic Optimization 169
    7.2.1 Uncertainty Representation 169
    7.2.2 FSO Unit Commitment 171
    7.2.3 Numerical Results for FSO UC 174
7.3 Robust Optimization 175
    7.3.1 Uncertainty Representation of Robust Optimization 176
    7.3.2 RO-Based Unit Commitment 177
    7.3.3 Numerical Result for RO-based UC 178
7.4 Chance-Constrained Programming 180
    7.4.1 CCP-Based Unit Commitment 181
    7.4.2 Conventional Generation Flexible Ramping Model 181
    7.4.3 Big-M Linearization Method 183
    7.4.4 Numerical Result for CCP UC 184
7.5 Conclusions 184
References 185

NOMENCLATURE

Decision variables at time t

$fr_{i,t(h)}^{\mathrm{up}}$:	Flexible ramping up of generator i
$frs_t^{\mathrm{up}}, frs_t^{\mathrm{down}}$:	Up or down flexible ramping surplus award
$g_{i,t}$:	Dispatch of generator i

DOI: 10.1201/9781003271857-7

$gr_{i,t(h)}^{\text{up}}, gr_{i,t(h)}^{\text{down}}$:	Up or down ramp rate limit of generator i
$\bar{l}_t^{\text{net}}$:	Expected net demand
$\overline{pb}_{v,t}^{\text{wind}}$:	Realization of wind (post curtailment) of level v
$pb_{s,t}^{\text{net}}$:	Realization of net demand of level s

Parameters

$g_i^{\text{max}}, g_i^{\text{min}}$	Max or min power of generator i
$g_{v,t}^{\text{wind}}$	Wind power realization of number v at time t
$\bar{l}_{b,t}$	Predicted number demand at bus b at time t
$l_{o,t}$	Realization of demand at number o at time t
$l_{s,t}^{\text{net}}$	Realization of net demand at number s at time t
$pb_{d,t}$	Realization of demand at number d
$pb_{v,t}^{\text{wind}}$	Realization of wind (precurtailment) at number v
sr_t	Spinning reserve at time t
$sd_t, su_{i,t}$	Shut-down or start-up cost of generator i at time t
$\text{time}_b^{\text{on}}, \text{time}_b^{\text{off}}$	Minimum on or off time of bus b
$\text{time}_i^{\text{on}}, \text{time}_i^{\text{off}}$	Minimum on or off time of generator i
$\varepsilon^{\text{up}}, \varepsilon^{\text{down}}$	Confidence level of upward/downward flexible ramping surplus

Decision variables (Binary)

$u_{i,\,t(h)}$	On or off status of generator i at time t
$y_{m,\,t},\, z_{s,\,t}$	Auxiliary variables
$\alpha_{s,\,t},\, \beta_{s,\,t}$	Auxiliary variable for upward/downward net demand ramp
$\gamma_{k,\,t}$	Auxiliary variable for expected net demand
$\omega_{s,\,t}$	Auxiliary variable for wind curtailment

Vectors

K_D, K_G	Bus-load, bus-generator incident vector
P^D, P^G	Demand vector, generation power dispatch vector
PL^{max}	Maximum power flow limit vector
SF	Shift factor (power transfer distribution factors)

7.1 STOCHASTIC OPTIMIZATION

Generally, optimization approaches have been utilized in power system operation for various problems, which includes economic dispatch, optimal power flow, and unit commitment (Wood, Wollenberg, and Sheblé 2013). The recent trend of introducing a considerable amount of renewable energy generation (REG) into the grid has not only caused the operation of modern electricity markets to be more complicated but also increased the need for load balancing. This has induced the requirement of short-term generation and nongeneration resources to maintain the power system flexibility, which will eventually lead to a more reliable system (Li et al. 2016). REG,

Role of Stochastic Optimization

particularly wind power, is different from traditional fossil fuel generation, as it cannot be forecasted precisely well in advance (*uncertainty*). Conversely, depending on meteorological conditions, REG varies with time (*variability*), especially when dealing with a shorter time step problem. Implementation of stochastic programming-based UC model, to cope with the uncertainty incorporated from the integration of REG, has been broadly reported in the literature, throughout the last few years (Tan, Shaaban, and Kadir 2019).

Ever since the beginning of UC, two waves of revolutions have been encountered, from the perspective of industrial and research applications. During the first wave of revolution, several deterministic UC methods were present including Lagrangian relaxation, priority list, dynamic programming, and Mixed Integer Linear Programming (MILP). Nonetheless, MILP has turned out to be the state-of-the-art solution method amongst them all (Guan, Zhai, and Papalexopoulos 2003, Ott 2010). On the other hand, in the second wave of revolution, the transformation of traditional deterministic methods to stochastic UC model was emphasized. This development was primarily driven by the evolving challenges to the operation of the grid over the past 30 years, which involves deregulation of power networks, integration of REG, and the surging requirement of power system flexibility, reliability, and resiliency.

UC is regularly used by power utility companies in deregulated energy markets, especially in day-ahead reliability assessment and real-time UC and market clearing, mainly for minimization of operation cost. In general, the UC problem is verified as nondeterministic polynomial-time hard (NP-hard) (Guan, Zhai, and Papalexopoulos 2003), wherein the size of the UC problem will cause the computation time to increase exponentially. This is mainly caused due to the generation ON/OFF status (units' commitment) which is formulated as binary variables and the integration of numerous operating constraints.

Several methods were suggested for the deterministic UC, for instance, the traditional priority list (Shoults et al. 1980, Lee 1991, Lee and Feng 1992, Senjyu et al. 2003) and dynamic programming (Snyder, Powell, and Rayburn 1987, Momoh 2008, Sen and Kothari 1998, Siu, Nash, and Shawwash 2001). These methods were under continued research and development over the years, until they were in full swing. Finally, due to the computation tractability and the elasticity in the setting of operating constraints, the most frequently used MILP methods managed to replace the aforementioned traditional methods. In a deterministic MILP-based UC model, the general formulation usually consists of a total production cost minimization objective function, which is formulated with the aim to minimize the shut-down and start-up costs, along with the fuel cost. The objective function also comes with several operating constraints (e.g., power balance, min/max generation, spinning reserve, min up/down time, generation unit ramping, along with power flow constraints). The latter was introduced and explained definitively by Padhy (2004) and Abujarad, Mustafa, and Jamian (2017).

Generally, grid flexibility is described as the capability of a grid to cope with the variability and uncertainty of demand and supply. Such designation comes in conjunction with the increasing penetration levels of the total contribution of electricity sourced from REG in the networks, which has significantly brought in uncertainty and variability into the power grid. Successful implementation of REG into the power systems leads to a new operation concept, more specifically, one that involves diverse organizational and technical challenges, along with unprecedented operation scenarios that entail a major coordination. These challenges could be solved with the use of technology-agnostic operational procedures and/or new market frameworks.

To enhance the planning and operation of power system with REG, several practices were proposed, for example, the deployment of flexible generation and non-generation resources (energy storage or demand response). Furthermore, with the aim to enhance power system reliability, a promising method, stochastic UC, was proposed, to replace current deterministic UC models. Stochastic programming can model the uncertainty via scenario generation, or integrate with the parameters or coefficients of uncertainty as continuous variables, which comes with a presumed distribution of probability. Even though in various research fields (Birge and Louveaux 2000), modeling of uncertainty with full stochastic optimization (FSO) is already a mature technology, its implementation to the NP-hard UC problem comes with several bottlenecks and barriers. Furthermore, Generation Companies (GENCOs) and Independent System Operators (ISOs) are worried about the complexity and practicability of FSO, to be implemented in electricity markets, mainly due to the computation intractability issue and the higher computational demand.

Several stochastic optimization methods were studied and proposed for solving UC in the literature, which can be categorized into three main methods: FSO, RO, and CCP. Figure 7.1 illustrates stochastic UC methods, as well as their corresponding model structures. The first step in stochastic optimization is the uncertainty modeling, which is essential and quite different among the three stochastic UC methods (Aien, Hajebrahimi, and Fotuhi-Firuzabad 2016). In general, there are three major ways to model the uncertainty: scenario generation, uncertainty sets, and chance constraints, for FSO, RO, and CCP, correspondingly, which will be further reviewed in the forthcoming sections. Moreover, various sources of uncertainties were modeled throughout the literature, which include the variation of demand, intermittency related to REG in power systems, and malfunctioning of power system components.

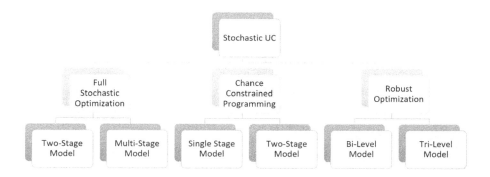

FIGURE 7.1 Categorization of stochastic UC models (Tan, Shaaban, and Kadir 2019)

7.2 POWER SYSTEM OPERATION WITH STOCHASTIC OPTIMIZATION

Stochastic optimization-based UC is known to be a promising tool to solve any REG uncertainty-related power system planning and operation problems. In order to effectively model uncertainties in the UC model, the elemental basic concept applied in the FSO is to generate sufficient scenario numbers to define the uncertainty. In contrast to the deterministic UC model with auxiliary reserve constraints, the FSO manages to further minimize the operating cost, along with the capability in enhancing the system reliability. Scenario-based uncertainty modeling is usually implemented in FSO, along with a two-stage FSO UC model, which will be further elaborated in the upcoming subsections.

7.2.1 UNCERTAINTY REPRESENTATION

In general, the most commonly implemented approach in stochastic optimization, particularly for FSO-based UC model, is the scenario-based uncertainty modeling method. Throughout the method, the uncertainty is denoted by numerous scenarios, in which each particular scenario signifies a potential realization of an uncertainty factor. Probability Density Function (PDF) is utilized to approximate the true probability distribution of a specific uncertainty, that is further utilized by the Monte-Carlo simulation, for scenario generation. Since a UC model usually depends on the forecasting algorithm to predict and generate future scenarios, a high-accuracy forecasting algorithm is essential to come out with an efficient FSO UC model.

Throughout the wind power scenario generation process, the temporal and spatial effects of wind power need to be considered. This is owing to the unavailability of the wind resources, when or where they are needed. Comprehensive validation of scenario quality and the modeling of forecast errors, which relate to various wind power scenario generation approaches, were discussed (Pierre et al. 2009, Morales, Mínguez, and Conejo 2010, Pinson and Girard 2012). Ortega-Vazquez and Kirschen (2009), Wang, Shahidehpour, and Li (2008), and Billinton et al. (2009) presented a FSO UC problem, where wind speed is modeled as a normal distribution PDF, while Monte-Carlo is utilized to produce a scenario. On the other hand, wind speed is represented by using a more accurate Weibull PDF (Venkatesh et al. 2008). Besides the wind generation, another REG resource is the solar energy, which also contains a significant level of uncertainty.

Various solar and wind generation scenarios are depicted in Figure 7.2a and b correspondingly. The figures demonstrate the discrepancy between solar and wind generation, in terms of the characteristics of the uncertainty. In particular, the spatial uncertainty of wind generation is obviously much significant, as compared to that of solar generation. This is mainly due to the diurnal pattern of solar generation that has a tendency to focus on the peak of the solar generation curve. Thus, the widely known California ISO duck curve is commonly caused by the high penetration level of solar generation. Undeniably, solar generation will result in steeper hourly ramp rates, which for sure will lead to a more strenuous impact on the power grids.

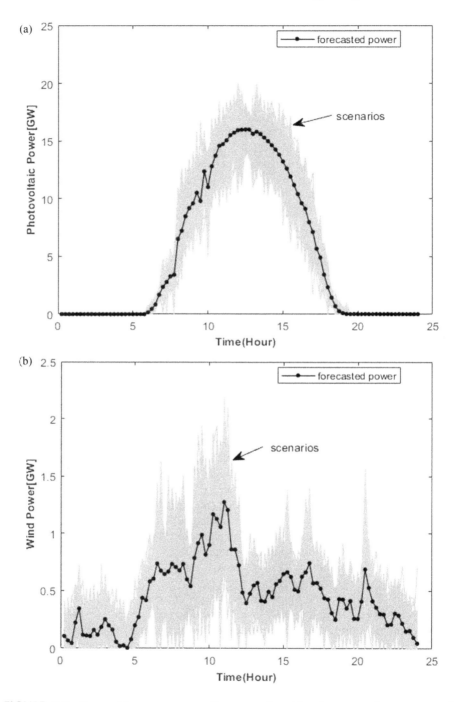

FIGURE 7.2 Renewable energy generation scenarios: (a) solar generation curves and (b) wind generation curves.

Role of Stochastic Optimization 171

Throughout the literature, two eminent challenges are observed, which restrict the practicality of the FSO model, in order for it to become the state-of-the-art stochastic UC model. These are as follows: (1) every scenario may not certainly incorporate the intermittency of the REG, as all the FSO scenarios are produced by Monte-Carlo in a random nature. (2) The capability of FSO model to handle problems with massive scenario numbers is limited, owing to the restriction of the current workstations' computational capability. Theoretically speaking, a substantial enhancement in the stochastic UC solution quality can be observed, by simply incorporating massive scenario cases. In general, a larger set of scenario cases is probable to attain a more comprehensive uncertainty modeling. Nonetheless, owing to the curse of dimensionality, as the set of scenario cases is enlarged more than a certain threshold value, only a minimal enhancement in the solution quality will be noticed. Additionally, computation requirements increase exponentially with the increase of scenario numbers, and thus, a compromise between accuracy and computational tractability of the FSO model, in that case, is inevitable. Subsequently, to overcome challenge (1), scenario reduction approaches were proposed, to decrease scenario numbers, with the aim of maintaining the solution accuracy. In general, the scenarios are clustered based on the distance between different scenarios, and then the comparable scenarios will be removed (Growe-Kuska, Heitsch, and Romisch 2003, Holger and Werner 2003, Sumaili et al. 2011). Nevertheless, certain percentage of the uncertainty information will be, inevitably, gone throughout the scenario reduction practice, which will cause an overall lower quality of FSO solutions.

7.2.2 FSO UNIT COMMITMENT

In general, a two-stage UC model is the most reported and formulated FSO model in the literature. Essentially, a two-stage FSO UC model usually deals with two stages of decision variables, that can be usually divided into day-ahead and intra-hour decision variables. The two-stage FSO UC can be formulated as

$$\min_{u \in U} c^T u + E_s\left[F(u,s)\right], \forall u \in U, s \in S \tag{7.1}$$

where u is the decision variable for day-ahead UC, typically consisting of base-load generation or traditional fossil fuel generation, for instance, coal and nuclear generations. Owing to the inability of the aforementioned base-load generation to rapidly shut-down and start-up in the intra-hour time frame, the day-ahead UC decision variables are solved in advance. Furthermore, U is termed as the decision variable set of traditional generation, which is subjected to technical operation constraints, for instance, minimum on or off time and ramp limit, etc. (Zheng et al. 2013). Vector c includes the generation shut-down and generation start-up costs, formulated as a parameter. In the second stage of Eq. (7.1), E_s is defined as the expected cost of intra-hour scheduling. Another parameter vector s is formulated, which comprises the scenario generated via the uncertainty modeling (introduced in the above subsection). s can be depending on either the discrete or continuous PDF. Nonetheless, to avoid the curse of dimensionality problem that is always linked to FSO, Monte-Carlo scenario generation is again utilized to convert the continuous PDF into discrete scenarios.

The FSO-based UC presented in this book chapter includes wind generation uncertainties, as well as the cost function (7.2), which includes traditional generation fuel cost, generation shut-down, and generation start-up costs. The day-ahead UC model can be expressed as

$$\min \sum_{s=1}^{S} P_s \sum_{i,t=1}^{I,T} \left[F_{c,its}(g_{i,t}) + su_{i,t} + sd_{i,t} \right] \tag{7.2}$$

s.t.

$$\sum_{i}^{I} g_{i,t} + g_t^{\text{wind}} = \sum_{b=1}^{B} l_{b,t} \tag{7.3}$$

$$\sum_{i=1}^{I} g_i^{\max} \cdot u_{i,t} \geq \sum_{b=1}^{B} l_{b,t} + sr_t \tag{7.4}$$

$$-PL^{\max} \leq SF \times \left[K_G \times P^G - K_D \times P^D \right] \leq PL^{\max} \tag{7.5}$$

$$g_{i,t+1} - g_{i,t} \leq gr_{i,t}^{\text{up}} \cdot u_{i,t} + g_i^{\min} \cdot \left(u_{i,t+1} - u_{i,t} \right) \tag{7.6}$$

$$g_{i,t} - g_{i,t+1} \leq gr_{i,t}^{dn} \cdot u_{i,t+1} + g_i^{\min} \cdot \left(u_{i,t} - u_{i,t+1} \right) \tag{7.7}$$

$$g_i^{\min} \cdot u_{i,t} \leq g_{i,t} \leq g_i^{\max} \cdot u_{i,t} \tag{7.8}$$

$$-u_{i,t-1} + u_{i,k} \leq 0, \forall 1 \leq k - (t-1) \leq \text{time}_i^{\text{on}} \tag{7.9}$$

$$u_{i,t-1} - u_{i,t} \leq 1, \forall 1 \leq k - (t-1) \leq \text{time}_i^{\text{off}} \tag{7.10}$$

The day-ahead FSO UC constraints listed above are denoted as follows:

- Equation (7.3) is the power balance constraint.
- Equation (7.4) entails system spinning reserve requirement.
- Equation (7.5) defines the transmission line limits.
- Equations (7.6) and (7.7) are the generation ramp limits.
- Equation (7.8) denotes maximum and minimum generation output limits.
- Equations (7.9) and (7.10) are the minimum on or off time limits of the generation.

Additionally, initial on/off states of generation units are also incorporated (Carrión and Arroyo 2006). To maintain the linearity, a piecewise linear fuel cost function is implemented, along with the stair-wise start-up cost function (Carrión and Arroyo 2006).

Owing to the sizable scenario numbers and variables in the FSO UC model, the Benders decomposition is usually applied to maintain the computation tractability, by decomposing the problem into two parts, which are known as master and sub

Role of Stochastic Optimization

problem. The day-ahead UC market generation commitment is decided without the consideration of wind uncertainty, in the master problem. However, additional operating constraints, such as the transmission network configuration, can be considered in the master problem. This indicates that the UC is solved by utilizing only the expected wind generation scenario. Next, Monte-Carlo generated scenarios are encompassed in the subproblem, representing intra-hour economic dispatch problem, to minimize operating costs triggered off by the divergence from the wind generation forecast error. As shown in Eq. (7.2), the expected total cost is the sum product of the operating cost of every subproblem and each scenario's probability of occurrence. Benders cuts are implemented to couple the master problem and subproblems (Birge and Louveaux 2000, Wang, Shahidehpour, and Li 2008, Wang et al. 2013, Aghaei et al. 2016), in which these cuts are back-fed to the master problem, in order to alter its solution, for further minimization of the subproblem objective function. The FSO UC simulation flowchart is illustrated in Figure 7.3.

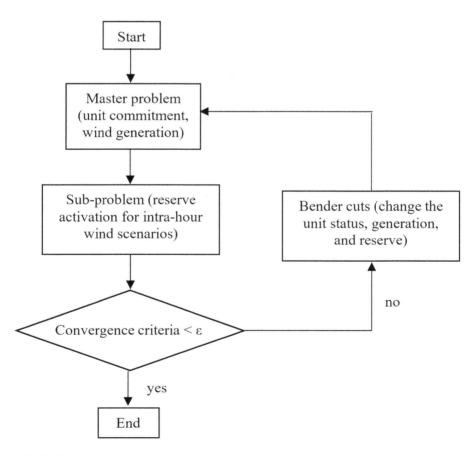

FIGURE 7.3 Bender decomposition for FSO UC problem.

7.2.3 NUMERICAL RESULTS FOR FSO UC

A modified IEEE 118-bus system is utilized for numerical simulation, which comprises 186 transmission lines, 118 buses, 33 thermal generators, and 4 wind farms included in buses 17, 27, 49, and 104 (http://motor.ece.iit.edu/Data/). The forecasts of wind generation up to 24 hours are shown in Figure 7.4. The resultant FSO UC MILP model is solved with IBM CPLEX 12.6 — a state-of-the-art solver for MILP problems.

As described in a previous subsection, a Monte-Carlo-based scenario reduction approach (Growe-Kuska, Heitsch, and Romisch 2003, Holger and Werner 2003, Sumaili et al. 2011) is utilized in the FSO UC model, with the aim to maintain the computation tractability. Scenario reduction combines the scenarios depending on a specific probabilistic metrics, for instance, the distance between scenarios. Two scenarios will be combined into one scenario, as long as their distance is smaller than a particular tolerance value. In this test case, up to 10,000 scenarios were generated and further decreased to 15 scenarios, as indicated in Figure 7.5. The FSO UC scheduling results with and without wind uncertainty are shown in Figure 7.6 (Nasrolahpour and Ghasemi 2015).

Case 1 is run without wind generation uncertainty, which means this is a typical deterministic UC model. The day-ahead UC total operation cost is $35,128,24. In this case, wind curtailment is not observed, since all wind farms are assumed to generate electricity based on the expected wind generation scenario. In contrast, Case 2 incorporates wind uncertainty and run with the FSO model with 15 scenarios. Results show that implementation of wind scenarios will lead to up to $3,132.58 of wind curtailment penalty cost. The total operation cost of $41,522.91 shows a notable hike as compared to Case 1, mainly due to the wind curtailment.

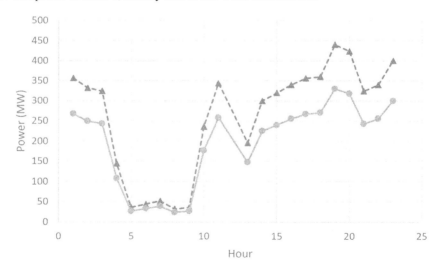

FIGURE 7.4 IEEE 118-bus system hourly wind power (Nasrolahpour and Ghasemi 2015).

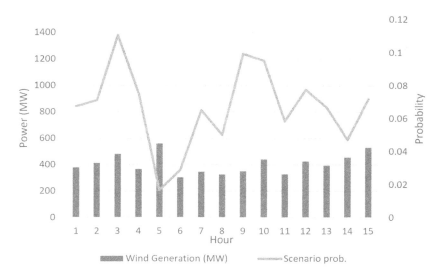

FIGURE 7.5 15 wind scenarios with probability at hour 23 in bus 17, IEEE 118-bus system (Nasrolahpour and Ghasemi 2015).

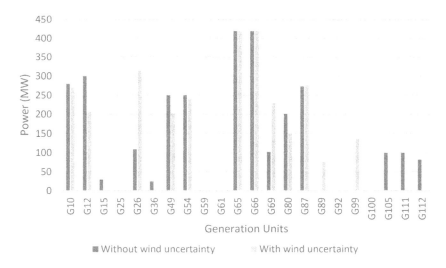

FIGURE 7.6 Scheduling without wind uncertainty and with wind uncertainty for 24-hour in IEEE 118-bus system (Nasrolahpour and Ghasemi 2015).

7.3 ROBUST OPTIMIZATION

In the past decade, RO has picked up an increasing interest in both academia research and industry applications, particularly in stochastic UC. As compared to FSO, RO-based UC tries to integrate REG uncertainty via uncertainty range, rather than depending on the PDFs and scenario generation methods. Moreover, standing apart from FSO UC models, which minimize the total expected cost, RO only minimizes

176   Renewable Energy Integration to the Grid

the worst-case scenario objective function, with the consideration of all possible solutions from the random realization within the uncertainty range (Jiang, Wang, and Guan 2012). In fact, this leads to the generation of over-conservative solutions, which is the main drawback of RO-based UC. However, RO is capable of maintaining the computationally tractability, even when solving middle and large-scale UC problems (Zheng, Wang, and Liu 2015).

7.3.1 Uncertainty Representation of Robust Optimization

As mentioned above, RO mainly implements the uncertainty range to model the prevalent system uncertainty. The most fundamental RO uncertainty range in UC models is the box interval uncertainty sets, which can be formulated as

$$\left[\max\left\{ 0, \bar{\xi} - z_\alpha \sigma \right\}, \bar{\xi} + z_\beta \sigma \right] \tag{7.11}$$

where z_α and z_β are the parameter probability distribution's α- and β-quantiles, correspondingly, where $\alpha < \beta$. $\bar{\xi}$ is the uncertain variable's mean value, and σ is the variance of the uncertain variable. Several historical data of demand are depicted in Figure 7.7a and b, for one week (24 hours × 7 days). Firstly, a single standard deviation value is set as the lower and upper bounds in Figure 7.7a. This is mainly to show that the respective boundary is not capable of covering the realizations for all the demand scenarios. If this setting is further implemented into RO-based UC, it will result in a risky solution, as there is a possibility of not meeting the demand at certain hour. The latter will require further rescheduling of the generation units. Nevertheless, in Figure 7.7b, the boundaries are increased to three standard deviation values, resulting in the expansion of the uncertainty range to trump up a larger boundary. This will lead to an over-conservative solution (overly protective and costly), as the RO-based model is solved for worst-case realization.

In this book chapter, the box interval of wind REG uncertainty for RO-based UC is modeled via choosing the cluster of uncertainty, Γ_b, which describes as the number of intervals with big deviations (three standard deviation value). It can also be self-defined by system operators. This uncertainty modeling can guarantee the RO solution to be viable with a high possibility under various realization scenarios of wind power. The RO uncertainty set can be formulated as

$$D := \left\{ w \in \mathbb{R}^{|B| \times T} : \sum_{t=1}^{T} \left(z_{bt}^+ + z_{bt}^- \right) \le \Gamma_b, \right.$$

$$\left. w_{bt} = W_{bt}^n + z_{bt}^+ W_{bt}^{\mathrm{up}} + z_{bt}^- W_{bt}^{\mathrm{low}}, \forall t, b \in B \right\} \tag{7.12}$$

where W_{bt}^n represents the forecasted wind generation, and W_{bt}^{up} W_{bt}^{low} are defined as the lower and upper boundaries of wind generation, respectively. Two binary variables z_{bt}^+ and z_{bt}^- are established, denoting that the wind power archives either the upper boundary $\left(z_{bt}^+ = 1 \right)$ or the lower boundary $\left(z_{bt}^- = 1 \right)$. If both binary variables are set to be zero, $\left(z_{bt}^+ = z_{bt}^- = 0 \right)$, then the wind power is assumed to meet its forecasted value.

Role of Stochastic Optimization

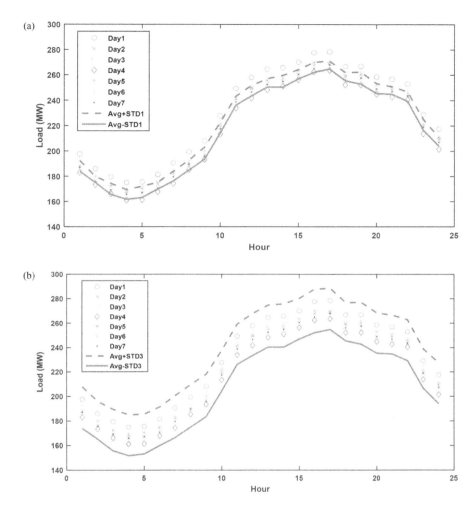

FIGURE 7.7 Demand distribution with confidence boundary: (a) with single standard deviation value and (b) with three standard deviation value.

7.3.2 RO-Based Unit Commitment

Throughout the literature, and similar to FSO UC model, RO-based UC primarily consists of a two-stage models. Based on the uncertainty set derived in Eq. (7.12), the RO-based UC problem formulation is expressed as

$$\min_{u \in U} \left\{ c^T u + \max \left[F_{v \in V}(u, v) \right] \right\}, \forall u \in U, v \in V \tag{7.13}$$

where c describes the shut-down and start-up costs, u and U are UC decision variables in the first stage, and feasible UC decision variable set, correspondingly, comparable to the FSO UC. Parameter vector v is an uncertain variable, and V is the range of uncertain variable set. Moreover, the objective function, in Eq. (7.13), has the connotation to minimize the most expensive operating cost for the worst-case realization, throughout the intra-hour time horizon. For instance, the most expensive realization can be formulated as

$$\min \sum_{i,t=1}^{I,T} \left[su_{i,t} + sd_{i,t} \right] + \max_{q \in D} \min \sum_{i,t=1}^{I,T} \left[F_{c,it}(g_{i,t}) \right] \quad (7.14)$$

s.t Eqs. (7.3)–(7.10)

7.3.3 Numerical Result for RO-based UC

For RO-based UC result simulation, the IEEE 118-bus system is again utilized, for consistency. Figure 7.8 illustrates the upper and lower limit values, along with the forecasted value of wind generation (Jiang, Wang, and Guan 2012). The resultant day-ahead RO-based UC MILP problem is solved with IBM CPLEX 12.6.

First, the system performance is tested under various uncertainty levels. The extent of uncertainty is allowed to be varied within an interval of [2, 10]. In contrast, the wind uncertainty level ratio is varied within an interval of [0, 0.5]. The numerical results are presented in Table 7.1, which also include the following:

a. Lower bound is attained by solving RO model without constraints.
b. Upper bound is attained via Monte-Carlo simulation of the wind generation.
c. Correspondingly, the optimal gap is estimated as *Optimal Gap = (UB − LB)/ LB × 100%*.

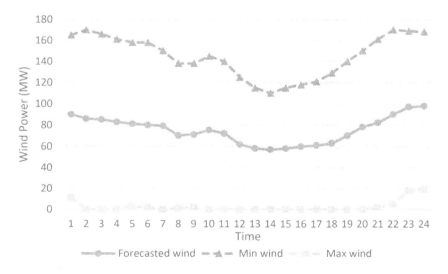

FIGURE 7.8 Wind generation over 24-hour horizon (Jiang, Wang, and Guan 2012).

TABLE 7.1

Numerical Results under Various Uncertainty Levels – Optimal Values and Worst-Case Performance for IEEE 118-Bus System (Jiang, Wang, and Guan 2012)

Operation cost ($)	Pump Storage Ratio								
	0.1			0.3			0.5		
	LB	UB	Opt. Gap (%)	LB	UB	Opt. Gap (%)	LB	UB	Opt. Gap (%)
$\Gamma = 2$	1,098,179	1,099,079	0.08	1,097,770	1,098,785	0.09	1,097,871	1,098,764	0.08
$\Gamma = 6$	1,139,800	1,140,891	0.10	1,139,997	1,141,030	0.09	1,139,796	1,140,837	0.09
$\Gamma = 10$	1,178,730	1,179,789	0.09	1,178,678	1,179,806	0.10	1,178,644	1,179,450	0.07

TABLE 7.2

Numerical Results of Optimal Values and Gaps, Under Various Wind Power Penetration Levels for IEEE 118-Bus System (Jiang, Wang, and Guan 2012)

Penetration Level		0.5	1.0	1.5
Operation cost ($)	LB	1,219,531	1,139,997	1,072,821
	UB	1,220,460	1,141,030	1,073,839
	Opt. Gap (%)	0.08	0.09	0.09

In the test cases, it is observed that statistically RO generates a viable solution for every timestep with ratio $\in [0, 0.5]$ and $\Gamma \in [2, 10]$. In addition, most optimality gaps are smaller than 0.1%, which demonstrates that the RO can provide a solution that is very efficient in obtaining near-optimal solutions. Then, the performance of 118-bus system subjected to several wind power penetration levels is examined. Numerical results for penetration level from 0.5 to 1.5 are displayed in Table 7.2. Overall, RO produces over-conservative results with high operation cost.

7.4 CHANCE-CONSTRAINED PROGRAMMING

Most of the stochastic programming techniques available in the literature, in particular the FSO, tend to minimize operation costs, while fulfilling all practical constraints under plausible system scenario(s). FSO may comprise a sizable number of extreme system scenarios. This, in turn, may well lead to prohibitively expensive solutions. Such apprehension prompted the notion of chance-constrained programming (CCP)-based UC models. Supplemental probabilistic constraints are encompassed in CCP UC models to further limit the risk probability of being exposed to an explicit set of unit commitment decisions. Various risk indices were put forward in stochastic UC models, such as the consideration of the Loss of Load Probability (LOLP) (Ozturk, Mazumdar, and Norman 2004, Zheng et al. 2013, Wang, Wang, and Guan 2011, Zhao et al. 2014) as well as Conditional-Value-at-Risk (CVaR) (Pozo and Contreras 2013, Huang, Zheng, and Wang 2014). The majority of CPP's uncertainty modeling is, however, quite comparable to the FSO, employing scenario-based uncertainties. Hence, an elaborate exposition of such application in CPP will not be replicated in this chapter.

As outlined previously, the essential idea of CCP is to account for the uncertainty information by appending probabilistic constraints into the original problem formulation. A generic CCP optimization problem can thus be stated as

$$\min_{u \in U} F(u,s), \forall u \in U, s \in S \tag{7.15}$$

s.t.

$$\Pr\left[A(u,s) \geq \bar{b}_{i,s} \right] \geq 1 - \varepsilon, \forall i, 1 \ldots, I \tag{7.16}$$

The cost function $F(u, s)$ is the same as the FSO UC decisions for the first stage, whereas U represents the region of feasible deterministic solutions (the variable

Role of Stochastic Optimization

set of feasible commitment decisions). Pr[·] symbolizes the likelihood of satisfying the inequality constraints, $\bar{b}$ is a variable vector, which acquires its values from the uncertainty vector s, and A is a constraint mapping parameter. The variable ε is a user-preset confidence level that ranges between 0 and 1 and is stipulating that the probability function $A(u, s) \geq \bar{b}_{i, s}$ is allowed to be violated within the predefined confidence level (Reich 2013).

Generally speaking, Eq. (7.15) is an *individual chance constraint.* In other words, a specific likelihood is disjointedly appointed to every inequality constraint. With such a formulation, different uncertainty parameters can be assigned to assorted confidence levels subject to system operator preferences. Numerous single CCP models that integrate the uncertainties in demand along with REG were reported in the literature. Reserve capabilities under wind uncertainty were modeled using single CCP (Restrepo and Galiana 2011, Wu et al. 2014, Ding et al. 2010, Wang et al. 2017). Furthermore, a single CCP model was formulated to minimize the risk of wind curtailment and warrant a higher utilization of wind power (Wang, Guan, and Wang 2012). In this book chapter, however, a single CCP is employed. The underlying purpose of such a choice is to ensure that the system has adequate flexible ramping requirement that makes it capable of meeting the uncertainties of demand as well as REG variability (Tan, Shaaban, and Abdullah 2017, Shaaban, Tan, and Abdullah 2018).

7.4.1 CCP-Based Unit Commitment

The contemplated CCP day-ahead scheduling employs the objective function (7.17). The aim is to minimize generation cost, both shut-down and start-up costs, along with the CCP auxiliary function μ:

$$\min \sum_{i,t=1}^{I,T} \left[F_{c,its}\left(g_{i,t}\right) + su_{i,t} + sd_{i,t} \right] + \mu \tag{7.17}$$

The above objective function is solved subject to Eqs. (7.3)–(7.10). The formulation incorporating the chance constraints is listed as follows.

7.4.2 Conventional Generation Flexible Ramping Model

The flexible ramping capacity of a conventional generation can be modeled through the ramp rate and operating bounds of the unit in the specified response time (Wu et al. 2015). The latter can be described as

$$fr_{i,t}^{\text{up}} = \min\left\{ \left[u_{i,t} \cdot \left(g_i^{\max} - g_{i,t} \right) \right], gr_{i,t}^{\text{up}} \right\} \tag{7.18}$$

$$fr_{i,t}^{\text{down}} = \min\left\{ \left[u_{i,t} \cdot \left(g_{i,t} - g_i^{\min} \right) \right], gr_{i,t}^{dn} \right\} \tag{7.19}$$

The net demand along with the corresponding net demand discrete levels is demonstrated in Figure 7.9a (Tan and Shaaban 2020). They are then reproduced as discrete

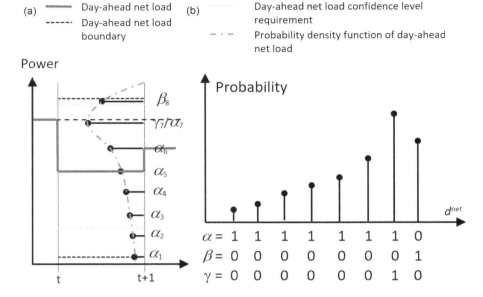

FIGURE 7.9 Flexible ramping surplus in the adopted chance-constrained rescheduling model: (a) net demand ramp and (b) net demand discrete PDF.

PDF in Figure 7.9b. α, β, and γ are three binary decision variable sets introduced in the formulation. γ is the expected net demand reference value. As the discrete level s is very close to the expected net demand, γ takes the value of 1. To further clarify such a notion, and looking back to Figure 7.9a, if the expected net demand is closest to the 7th γ, in this case, γ can be mathematically expressed as [0 0 0 0 0 0 1 0]. On the other hand, α and β turn to be 1 at downward and upward flexible ramping regions correspondingly, as illustrated in Figure 7.9b. The flexible ramping model can be listed as

$$\min\left\{\mu = \sum_{t=1}^{T}\sum_{s=1}^{S} \gamma_{s,t} \cdot l_{s,t}^{net}\right\} \tag{7.20}$$

$$\sum_{s=1}^{S} \gamma_{s,t} = 1 \tag{7.21}$$

$$\sum_{s=1}^{S} \gamma_{s,t} \cdot l_{s,t}^{net} \geq l_{t}^{-net} \tag{7.22}$$

$$\beta_{s,t} \geq \gamma_{s,t} \tag{7.23}$$

$$\beta_{s-1,t} \leq 1 - \gamma_{s,t} \tag{7.24}$$

Role of Stochastic Optimization

$$\beta_{s,t} \geq \beta_{s-1,t} \qquad (7.25)$$

$$\alpha_{s,t} = 1 - \beta_{s,t} \qquad (7.26)$$

The objective function incorporates Eq. (7.20) as an auxiliary objective function μ. In general, minimizing μ entails minimizing the mean value of net demand. This is tantamount to maximizing wind generation. Such an objective is in tandem with the day-ahead UC objective that minimizes the total costs, while maintaining the reduction of REG to a minimum.

The chance constraints, including the wind curtailment effect (Restrepo and Galiana 2011), for the day-ahead UC model with flexible ramping procurement are cast as

$$pb\left(\sum_{i,t=1}^{IT} \left[fr_{i,t}^{\text{up}} \right] \geq l_{s,t+1}^{\text{net}} - l_t^{-\text{net}} \right) \geq 1 - \varepsilon^{\text{up}} \cdot \sum_{s}^{S} pb_{s,t+1}^{\text{net}} \cdot \beta_{s,t} \qquad (7.27)$$

$$pb\left(\sum_{i,t=1}^{IT} \left[fr_{i,t}^{\text{down}} \right] \geq l_t^{-\text{net}} - l_{s,t+1}^{\text{net}} \right) \geq 1 - \varepsilon^{\text{down}} \cdot \sum_{s}^{S} pb_{s,t+1}^{\text{net}} \cdot \alpha_{s,t} \qquad (7.28)$$

According to Eqs. (7.27)–(7.28), the upward/downward flexible ramping are apportioned to match the uncertainty flexible ramping requirement at time t. This will be carried out up to a predetermined confidence level, $\varepsilon^{(\text{up, down})}$. The sum of upward/downward net demand uncertainty is then multiplied with the predetermined confidence level, $\varepsilon^{(\text{up, down})}$, respectively. If ε^{up} is set to 0.4, for example, it implies that all the downward uncertainty and 60% of upward uncertainty are met.

7.4.3 Big-M Linearization Method

The inherent nonlinearity of the chance constraints makes their solution complicated. The conventional Big-M method converts (7.27)–(7.28) into a set of linear equivalents (Restrepo and Galiana 2011). To carry out the linearization, binary variables, $Z_{s,t}^{\text{up}}$ and $Z_{s,t}^{dn}$, are introduced. Every binary variable is associated to a single realization of net demand $l_{s,t}^{\text{net}}$. This can be expressed as

$$\sum_{i,j,b=1}^{I,J,B} \left[fr_{i,t}^{\text{up}} + fr_{j,t}^{\text{up}} + fr_{b,t}^{\text{up}} \right] + l_t^{-\text{net}} - l_{s,t+1}^{\text{net}} + z_{s,t}^{\text{up}}, M \geq 0 \qquad (7.29)$$

$$\sum_{i,j,b=1}^{I,J,B} \left[fr_{i,t}^{\text{down}} + fr_{j,t}^{\text{down}} + fr_{b,t}^{\text{down}} \right] - l_t^{-\text{net}} + l_{s,t+1}^{\text{net}} + z_{s,t}^{dn}, M \geq 0 \qquad (7.30)$$

$$\sum_{s=1}^{S} Z_{s,t}^{(\text{up,down})} \cdot pb_{s,t}^{\text{net}} \leq \varepsilon^{(\text{up,down})} \cdot \sum_{s}^{S} pb_{s,t+1}^{\text{net}} \cdot \left(\alpha_{s,t}, \beta_{s,t} \right) \qquad (7.31)$$

184 Renewable Energy Integration to the Grid

With the Big-M method solving the preceding MILP model (7.17), (7.3)–(7.10), (7.18)–(7.26), and (7.29)–(7.31), a complete characterization of the stochastic CCP day-ahead UC problem can be realized. The IBM CPLEX 12.6 is again implemented to solve the MILP problem.

7.4.4 Numerical Result for CCP UC

To adjudge the efficacy of the proposed CCP-based UC model, it is implemented to the IEEE 118-bus system again. The testbed model has four 675 MW of wind turbines. With a wind penetration level amounting to 30.1%, wind turbines are disbursed at buses 15, 24, 54, and 96. The optimality gap is taken as 1.5% for all the case studies of the day-ahead UC model.

The base case, without wind generation, has an operating cost of $852,170. Numerical results with wind uncertainty, which includes the full-fledged wind contribution of 30.1%, are enumerated in Table 7.3. The result demonstrates that the incorporation of wind power renders a considerable decrement in production cost.

With the CCP confidence level increasing to 60% and 95%, signifying an increase in flexible ramping capacity demand, production cost and value of stochastic solution (VSS) are raised consequently. This is because the opportunity cost and wind curtailment are affected. When less expensive generation is cut back to supply for the upward flexible ramping or more costly generation is bumped up to supply for the downward flexible ramping, the opportunity cost component emerges. As a result, traditional generation units are subjected to supplementary ramping, to provide adequate flexible ramping capacity at every hour or intra-hour timestep. Along with the increase of flexible ramping capacity, $CVaR_{95}$ has decreased as well. This subsequently indicates that the number of load-shedding instances is diminished.

7.5 CONCLUSIONS

The uncertainties associated with variable REG combined with the trend to integrate large proportions of REG into the electricity grid system have sparked a lot of interest in using stochastic approaches in power system optimization. An exhaustive description of the full spectrum of power system operation problems and decision

TABLE 7.3
Numerical Results for Day-Ahead CCP-Based UC for IEEE 118-Bus System (Tan and Shaaban 2020)

Cases	Confidence Level (%)	Operating Cost ($)	Flexible Ramping Surplus Cost ($)	$CVaR_{95}$ (MW)	Wind Curtailment (%)	Total Cost ($)
Without wind	-	852,170	-	-	-	852,942
With wind	0	584,466	5,994	-	0.63	591,144
	60	608,162	4,702	156.89	3.91	614,186
	95	613,627	4,724	50.53	4.96	619,589

Role of Stochastic Optimization

support tools that are amenable to the use of stochastic optimization techniques cannot be compounded into a single book chapter, without violating the space limitation or missing out on application intricacies. This book chapter has, therefore, embarked on discussing the significance of state-of-the-art stochastic optimization approaches for system operation and decision-making, with a confined scope on the UC problems. A variety of modeling aspects were considered to facilitate the hosting of the rising levels of REG into the power network. Stochastic UC models, which include uncertainty modeling, were described. Stochastic optimization methods such as FSO, RO, and CCP were thoroughly modeled. The three stochastic optimization methods were established on the IEEE 118-bus system. Numerical results of the case study simulations applied on the same testbed system were presented and contrasted. These test results, corroborated by a critical review of the literature, show that FSO models produce a substantial problem size, also known as the curse of dimensionality, with hefty computational requirements. In contrast to RO models which may yield overly conservative results, the CCP UC model allows for greater flexibility in determining the reliability. This can be realized since a user-preset confidence level is used for the solution.

The strengths and weaknesses of each of the stochastic optimization methods, namely FSO, RO, and CPP, along with modeling and implementation details were the prime focus of this chapter. Nevertheless, some of these shortcomings can be circumvented by hybridizing FSO/RO and CCP/RO models to wind up with orderly computational solutions and unrestrictive results.

REFERENCES

Abujarad, S. Y., M. W. Mustafa, and J. J. Jamian. 2017. Recent approaches of unit commitment in the presence of intermittent renewable energy resources: A review. *Renew. Sustain. Energy Rev.* 70:215–223.

Aghaei, J., A. Nikoobakht, P. Siano, M. Nayeripour, A. Heidari, and M. Mardaneh. 2016. Exploring the reliability effects on the short term AC security-constrained unit commitment: A stochastic evaluation. *Energy* 114:1016–1032.

Aien, M., A. Hajebrahimi, and M. Fotuhi-Firuzabad. 2016. A comprehensive review on uncertainty modeling techniques in power system studies. *Renew. Sustain. Energy Rev.* 57:1077–1089.

Billinton, R., B. Karki, R. Karki, and G. Ramakrishna. 2009. Unit commitment risk Analysis of wind integrated power systems. *IEEE Trans. Power Syst.* 24 (2):930–939.

Birge, J., and F. Louveaux. 2000. *Introduction to Stochastic Programming*. Springer, New York.

Carrión, M., and J. M. Arroyo. 2006. A computationally efficient mixed-integer linear formulation for the thermal unit commitment problem. *IEEE Trans. Power Syst.* 21 (3):1371–1378.

Ding, X. Y., W.-J. Lee, J. X. Wang, and L. Liu. 2010. Studies on stochastic unit commitment formulation with flexible generating units. *Electr. Power Syst. Res.* 80 (1):130–141.

Growe-Kuska, N., H. Heitsch, and W. Romisch. 2003. Scenario reduction and scenario tree construction for power management problems. *IEEE Bologna Power Tech Conf. Proc.*, 23–26 June 2003.

Guan, X. H., Q. Z. Zhai, and A. Papalexopoulos. 2003. Optimization based methods for unit commitment: Lagrangian relaxation versus general mixed integer programming. *Proc. IEEE Power Eng. Soc. General Meeting*, 13–17 July 2003.

Holger, H., and R. Werner. 2003. Scenario reduction algorithms in stochastic programming. *Comput. Optim. Appl.* 24 (2):187–206.

Illinois Institute of Technology Power Group Test Case. Accessed 18 June 2021. http://motor.ece.iit.edu/Data/.

Huang, Y. P., Q. P. Zheng, and J. H. Wang. 2014. Two-stage stochastic unit commitment model including non-generation resources with conditional value-at-risk constraints. *Electr. Power Syst. Res.* 116:427–438.

Jiang, R., J. Wang, and Y. Guan. 2012. Robust unit commitment with wind power and pumped storage hydro. *IEEE Trans. Power Syst.* 27 (2):800–810.

Lee, F. N. 1991. The application of commitment utilization factor (CUF) to thermal unit commitment. *IEEE Trans. Power Syst.* 6 (2):691–698.

Lee, F. N., and Q. Feng. 1992. Multi-area unit commitment. *IEEE Trans. Power Syst.* 7 (2):591–599.

Li, N., C. Uçkun, E. M. Constantinescu, J. R. Birge, K. W. Hedman, and A. Botterud. 2016. Flexible operation of batteries in power system scheduling with renewable energy. *IEEE Trans. Sustain. Energy* 7 (2):685–696.

Momoh, J. A. 2008. *Electric Power System Applications of Optimization*, Second Edition. Taylor & Francis, Boca Raton, FL.

Morales, J. M., R. Mínguez, and A. J. Conejo. 2010. A methodology to generate statistically dependent wind speed scenarios. *Appl. Energy* 87 (3):843–855.

Nasrolahpour, E., and H. Ghasemi. 2015. A stochastic security constrained unit commitment model for reconfigurable networks with high wind power penetration. *Electr. Power Syst. Res.* 121:341–350.

Ortega-Vazquez, M. A., and D. S. Kirschen. 2009. Estimating the spinning reserve requirements in systems with significant wind power generation penetration. *IEEE Trans. Power Syst.* 24 (1):114–124.

Ott, A. L. 2010. Evolution of computing requirements in the PJM market: Past and future. *IEEE PES General Meeting*, 25–29 July 2010.

Ozturk, U. A., M. Mazumdar, and B. A. Norman. 2004. A solution to the stochastic unit commitment problem using chance constrained programming. *IEEE Trans. Power Syst.* 19 (3):1589–1598.

Padhy, N. P. 2004. Unit commitment-a bibliographical survey. *IEEE Trans. Power Syst.* 19 (2):1196–1205.

Pierre, P., M. Henrik, N. Henrik Aa, P. George, and K. Bernd. 2009. From probabilistic forecasts to statistical scenarios of short-term wind power production. *Wind Energy* 12 (1):51–62.

Pinson, P., and R. Girard. 2012. Evaluating the quality of scenarios of short-term wind power generation. *Appl. Energy* 96:12–20.

Pozo, D., and J. Contreras. 2013. A chance-constrained unit commitment with an n-K security criterion and significant wind generation. *IEEE Trans. Power Syst.* 28 (3):2842–2851.

Reich, D. 2013. A linear programming approach for linear programs with probabilistic constraints. *Eur. J. Oper. Res.* 230 (3):487–494.

Restrepo, J. F., and F. D. Galiana. 2011. Assessing the yearly impact of wind power through a new hybrid deterministic/stochastic unit commitment. *IEEE Trans. Power Syst.* 26 (1):401–410.

Sen, S., and D. P. Kothari. 1998. Evaluation of benefit of inter-area energy exchange of the Indian power system based on multi-area unit commitment approach. *Electr. Mach. Power Syst.* 26 (8):801–813.

Senjyu, T., K. Shimabukuro, K. Uezato, and T. Funabashi. 2003. A fast technique for unit commitment problem by extended priority list. *IEEE Trans. Power Syst.* 18 (2):882–888.

Shaaban, M., W.-S. Tan, and M. P. Abdullah. 2018. A multi-timescale hybrid stochastic/deterministic generation scheduling framework with flexiramp and cycliramp costs. *Int. J. Electr. Power Energy Syst.* 99:585–593.

Shoults, R. R., S. K. Chang, S. Helmick, and W. M. Grady. 1980. A practical approach to unit commitment, economic dispatch and savings allocation for multiple-area pool operation with import/export constraints. *IEEE Trans. Power App. Syst.* PAS-99 (2):625–635.

Siu, T. K., G. A. Nash, and Z. K. Shawwash. 2001. A practical hydro, dynamic unit commitment and loading model. *IEEE Trans. Power Syst.* 16 (2):301–306.

Snyder, W. L., H. D. Powell, and J. C. Rayburn. 1987. Dynamic programming approach to unit commitment. *IEEE Trans. Power Syst.* 2 (2):339–348.

Sumaili, J., H. Keko, V. Miranda, A. Botterud, and J. Wang. 2011. Clustering-based wind power scenario reduction technique. *Proc. of 17th Power Syst. Computation Conf.*, Stockholm, Sweden, 22–26 August 2011.

Tan, W. S., and M. Shaaban. 2020. Dual-timescale generation scheduling with nondeterministic flexiramp including demand response and energy storage. *Electr. Power Syst. Res.* 189:106821.

Tan, W. S., M. Shaaban, and M. P. Abdullah. 2017. Chance-constrained programming for day-ahead scheduling of variable wind power amongst conventional generation mix and energy storage. *IET Renew. Power Gen.* 11 (14):1785–1793.

Tan, W. S., M. Shaaban, and M. Z. A. Ab Kadir. 2019. Stochastic generation scheduling with variable renewable generation: Methods, applications, and future trends. *IET Gener. Transm. Distrib.* 13 (9):1467–1480.

Venkatesh, B., P. Yu, H. B. Gooi, and D. Choling. 2008. Fuzzy MILP unit commitment incorporating wind generators. *IEEE Trans. Power Syst.* 23 (4):1738–1746.

Wang, B., X. Yang, T. Short, and S. Yang. 2017. Chance constrained unit commitment considering comprehensive modelling of demand response resources. *IET Renew. Power Gen.* 11 (4):490–500.

Wang, J. D., J. H. Wang, C. Liu, and J. P. Ruiz. 2013. Stochastic unit commitment with sub-hourly dispatch constraints. *Appl. Energy* 105:418–422.

Wang, J., M. Shahidehpour, and Z. Li. 2008. Security-constrained unit commitment with volatile wind power generation. *IEEE Trans. Power Syst.* 23 (3):1319–1327.

Wang, Q. F., Y. P. Guan, and J. H. Wang. 2012. A chance-constrained two-stage stochastic program for unit commitment with uncertain wind power output. *IEEE Trans. Power Syst.* 27 (1):206–215.

Wang, Q., J. Wang, and Y. Guan. 2011. Wind power bidding based on chance-constrained optimization. *Proc. IEEE Power and Energy Soc. Gen. Meeting*, 24–29 July 2011.

Wood, A. J., B. F. Wollenberg, and G. B. Sheblé. 2013. *Power Generation, Operation, and Control*. Wiley, Hoboken, NJ.

Wu, H., M. Shahidehpour, A. Alabdulwahab, and A. Abusorrah. 2015. Thermal generation flexibility with ramping costs and hourly demand response in stochastic security-constrained scheduling of variable energy sources. *IEEE Trans. Power Syst.* 30 (6):2955–2964.

Wu, H., M. Shahidehpour, Z. Li, and W. Tian. 2014. Chance-constrained day-ahead scheduling in stochastic power system operation. *IEEE Trans. Power Syst.* 29 (4):1583–1591.

Zhao, C., Q. Wang, J. Wang, and Y. Guan. 2014. Expected value and chance constrained stochastic unit commitment ensuring wind power utilization. *IEEE Trans. Power Syst.* 29 (6):2696–2705.

Zheng, Q. P., J. H. Wang, P. M. Pardalos, and Y. P. Guan. 2013. A decomposition approach to the two-stage stochastic unit commitment problem. *Ann. Oper. Res.* 210 (1):387–410.

Zheng, Q. P., J. Wang, and A. L. Liu. 2015. Stochastic optimization for unit commitment: A review. *IEEE Trans. Power Syst.* 30 (4):1913–1924.

8 Dependence Modeling of Multisite Renewable Generations

B Rajanarayan Prusty
Vellore Institute of Technology, Vellore

Satyabrata Das
NIST Rourkela

CONTENTS

8.1 Introduction 189
8.2 Copula-Based Dependence Modeling for PSSA 190
    8.2.1 Parametric Family of Copula 192
        8.2.1.1 Comparison of Various Copula Functions........................ 192
    8.2.2 Multivariate Dependence Modeling 192
8.3 Result Analysis 193
    8.3.1 Selecting Appropriate Copula Function 193
    8.3.2 Modeling of Multivariate Dependence Structure Using
        Pair-Copula Function.. 194
8.4 Conclusion 196
References... 196

8.1 INTRODUCTION

There is an increasing interest in the adoption of probabilistic methods for power system analysis (Prusty and Jena 2017). The obtained probabilistic information of power system variables helps power engineers (planners and operators) with enough confidence for making realistic planning and operational decisions (Prusty and Jena 2018). The steady-state analysis is frequently used in power systems and is commonly known as the load flow analysis. For a renewable energy reach power system, inputs for such an analysis include load real and reactive powers, renewable generations, etc. (Fan et al. 2012, Prusty and Jena 2016). Probabilistic steady-state analysis (PSSA) refers to the adaptation of probabilistic load flow (PLF) to address the uncertainties of inputs for characterizing the uncertainties in the power system variables such as bus voltage magnitudes, real and reactive branch power flows, branch power losses, etc. The requirements for PSSA are threefold: uncertainty and dependence modeling, power system model development, and application of an uncertainty handling

DOI: 10.1201/9781003271857-8

method. In PSSA, power system inputs are referred to as random variables (RVs). Dependence modeling is one of the essential steps for PSSA. The dependencies among input RVs are no longer linear, and it is necessary to appropriately characterize the accompanying dependence structure(s) among input RVs. This chapter aims to summarize/compare various dependence modeling mechanisms and elaborate modeling steps for nonlinear dependencies among power system input RVs to hedge realistic decisions in system planning and operation through PSSA.

Accurate PSSA analysis demands accurate modeling of dependencies among input RVs (Usaola 2010, Gupta et al. 2014). Linear dependence modeling is frequently considered in PSSA owing to the difficulty in modeling precise dependence structures. Pearson product-moment correlation coefficient (PMCC) is used for the estimation of linear dependence (correlation). PMCC is calculated in the actual domain; therefore, it is affected by the marginal distributions. It is invariant under linear transformation and suitable for Gaussian and elliptically distributed RVs. PMCC calculation between two RVs is nonideal for a dependence measure of heavily-tailed marginal distributions where variance appears infinite (Prusty and Jena 2019). Further, it is nonideal to measure dependence structures such as tail and symmetric dependence, etc. It cannot measure the degree of nonlinear dependence. On the other hand, rank correlation is independent on the marginal distribution, and it always exists. It can take any value in the interval $[-1, 1]$. It is invariant under monotonic increasing transformations of input RVs. Spearman rank-order correlation coefficient (SRCC) and Kendall rank correlation coefficient (KRC) measure the monotonic relationship between input RVs and are more suitable to quantify different dependence structures (Wu et al. 2015). In addition to this, both fit a suitable Copula to input data. Hence, a measure of dependence considering both degrees of dependence and dependence structure is of greater interest. Initially, research on PLF had only considered linear dependence among input RVs which completely ignores nonlinear dependence and dependence structures. Disadvantages of such traditional methods are as follows: (1) all values of negative correlation cannot be modeled; (2) in case of the value of correlation of practical interest, PMCC matrix may not be positive definite; (3) although positive correlation can be accurately incorporated among input RVs exhibiting a parametric distribution, inaccuracy persists in case of nonparametric distribution of input RVs; and (4) traditional methods fail to incorporate nonlinear dependence among the input RVs. Implementation of the Copula theory overcomes the above pitfalls. A Copula function, otherwise known as a joint function, is referred to as the cumulative distribution function (CDF) of uniformly distributed RVs.

8.2 COPULA-BASED DEPENDENCE MODELING FOR PSSA

Dependence modeling mainly separates the influence of marginal distributions from the dependence structure (Wang et al. 2016). In practice, the dependence among input RVs in any probabilistic analysis is rarely linear. Modeling nonlinear dependencies associated with several input RVs adds additional complexity to the overall modeling framework. Most of the existing probabilistic models have solely focused on the randomness of load power and renewable generations but have not

Dependence Modeling

considered complex nonlinear dependencies. Hence, these models are not adequate when PSSA is performed for planning and operation. A more flexible Copula structure is required to capture various dependence patterns of input RVs. Ignorance of nonlinear dependence will cause significant errors in the analyses. The use of the Copula function can handle nonlinear dependence. A Copula is also referred to as a multivariate probability distribution with uniform marginals. Any multivariate distribution function can be expressed by a Copula joining its marginals. Copula differs not so much in the degree of association it provides but instead in which part of the association's distribution is vital. The conventional PMCC measures the overall strength of correlation between the input RVs but gives no information about how that varies across the distribution. In this chapter, the choice of a suitable Copula to model nonlinear dependence among input RVs of different dependence structure are discussed.

Sklar's theorem bridge between Copula and joint distribution of input RVs. Two RVs X and Y with CDFs F_X and F_Y are joined by Copula C if their joint distribution is expressed as

$$F_{XY}(x,y) = C\big(F_X(x), F_Y(y)\big) \tag{8.1}$$

If F_X and F_Y are continuous, then C is unique. Let $F_X(x) = u$ and $F_Y(y) = v$ are the realizations of the uniform RVs U and V, respectively. In this case, (8.1) can be rewritten as

$$C_{UV}(u,v) = F\big(F_X^{-1}(u), F_Y^{-1}(v)\big) \tag{8.2}$$

Copula functions can be easily constructed from the respective multivariate distribution functions with the help of (8.2). Copula density is defined by

$$c\big(F_X^{-1}(u), F_Y^{-1}(v)\big) = \frac{\partial^n F\big(F_X^{-1}(u), F_Y^{-1}(v)\big)}{\partial u \, \partial v} \tag{8.3}$$

Consequently, the joint density f can be expressed by

$$f\big(F_X^{-1}(u), F_Y^{-1}(v)\big) = c(u,v) f_1\big(F_X^{-1}(u)\big) f_2\big(F_Y^{-1}(v)\big) \tag{8.4}$$

where f_1 and f_2 are marginal densities, respectively.

Eqs. (8.1)–(8.4) can be easily extended to multivariate case in a similar fashion. When the dependence is measured in the actual domain, it is affected by the marginal distributions. In order to avoid this effect, stochastic dependence is modeled in the rank domain. The dependence function C is then used among ranks of the RVs. The complexity of using Copula is strictly related to the number of input RVs. Archimedean family is one of the most commonly used Copula families which include Gumbel Copula, Clayton Copula, Frank Copula, etc. Another important family is the elliptical Copula family. It includes Gaussian Copula, Student's t-Copula, etc. (Wang et al. 2016).

8.2.1 Parametric Family of Copula

Implicit Copula is derived from the well-known multivariate probability density function (PDF). They do not have simple closed forms, e.g., Gaussian Copula and t-Copula. Explicit Copula/Archimedean Copula, on the other hand, is not derived from multivariate PDF but does have simple closed forms, e.g., Clayton Copula, Gumbel Copula, and Frank Copula. Clayton and Gumbel Copula permit only positive dependence, whereas Frank Copula allows both negative and positive dependencies. The t-Copula and Archimedean family of Copulas can represent tail dependence. The Archimedean family of Copulas cannot be easily used for dimensions greater than two. Though Gaussian Copula can be used in this context, it is inadequate to capture all patterns of dependence.

8.2.1.1 Comparison of Various Copula Functions

The Gaussian Copula is relatively easy to sample, and different levels of correlation between the marginal can be easily determined. But it fails to represent dependence between the extreme values (tail dependence) of the input RVs. Further, it demands that input RVs must be continuous with Gaussian marginal. Student's t-Copula allows for some flexibility in covariance structure and tail dependence. Clayton Copula is suitable for lower tail dependence. It has a good range of exchangeable negative dependence compared with multivariate Gaussian distribution. It cannot account for negative dependence, and zero density exists in a particular region. It cannot correctly capture higher negative dependence, especially in the presence of a lower dimension. Gumbel Copula can capture strong upper tail dependence and weak lower tail dependence. A higher probability concentration appears in the right tail. Frank Copula has the flexibility in allowing the maximum range of dependence.

8.2.2 Multivariate Dependence Modeling

The nonlinear dependencies can be modeled using different Copulas such as (1) pair-Copula construction (PCC) and (2) nested Archimedean Copula (NAC). This chapter discusses the modeling of multivariate dependence using PCC. Many research articles have been published in the existing literature dealing with the dependence modeling of different data sets. Each proposed methodology has its significances and drawbacks. A comprehensive review of the literature using the Copula theory for dependence modeling was done. Pair-Copula methodology applied to multivariate financial data set to model the complex dependence patterns is presented in a study by Aas et al. (2009). Modeling spatial dependence using the pair-Copula theory is applied to multivariate wind data set in a study by Lu et al. (2014). Comparison of two models that differ in the representation of the dependence structure, namely the NAC and PCC, is highlighted in the study by Aas and Berg (2013). Wu et al. (2015) proposed a versatile probability model of PV generation based on PCC. The basic theory concerning these functions for dependence modeling and a brief idea about the Copula function are available (Papaefthymiou and Kurowicka 2009).

Dependence Modeling

8.3 RESULT ANALYSIS

This section discusses the results obtained using the Copula-based probabilistic PV generation multivariate dependence modeling. The programming for the simulation is done using MATLAB. The study uses the historical PV generation samples collected for 5 years. The PV generation data are collected for years, i.e., from 2012 to 2016, from three rooftop PV installations in the USA (refer to Figure 8.1), excluding leap days (PV Generation Data, 2014). Preprocessing of data is an essential step before uncertainty modeling. Therefore, the PV generation data are preprocessed using the improved sliding window prediction-based method (Jain, et al. 2021, Ranjan, et al. 2021) before nonlinear dependence modeling using Copula theory.

The analysis of results is done to model the nonlinear dependencies among PV generation data using Copula functions. The simulation is carried out at a time-instant (i.e., at noon) which is arbitrarily chosen for the analysis. The main objectives of the analysis include the following:

1. To choose an appropriate Copula function.
2. To model the multivariate dependence using a suitable pair-Copula function.

8.3.1 Selecting Appropriate Copula Function

The modeling approach selects the best Copula function among the families of Copula which can be done by estimating parameters arbitrarily of chosen Copula or by applying the goodness-of-fit test. For the dependence modeling using higher sample size, it is possible to identify the correct Copula function. For this purpose,

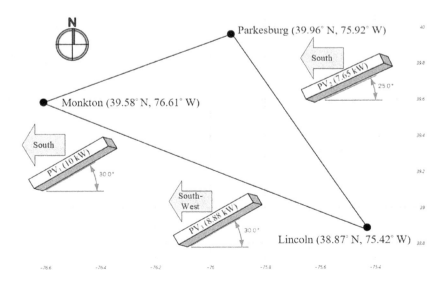

FIGURE 8.1 Details of PV arrays.

TABLE 8.1
Copula Parameters of the Paired PV Arrays

	Parameters	
Paired PV Arrays	**Gumbel (α)**	**Gaussian (θ)**
PV_1 and PV_2	1.9894	0.7041
PV_2 and PV_3	1.6434	0.5769
PV_1 and PV_3	1.5774	0.5438

estimating parameters of an arbitrarily chosen Copula is a common way for selecting a Copula that fits better than the other candidate Copula functions. Table 8.1 shows their corresponding Copula parameters. However, there is no certain way to choose a Copula that fits data precisely. Therefore, Gaussian and Gumbel are candidate Copulas for selection.

8.3.2 MODELING OF MULTIVARIATE DEPENDENCE STRUCTURE USING PAIR-COPULA FUNCTION

Table 8.2 shows the PMCC matrix among PV generations. The pair of PV arrays that possess the strongest correlation matrix is considered the root node in the first layer of PCC. Thus, based on the size of the PMCC matrix, the sequence of the RVs is determined. By applying marginal CDF transformation, predicted errors of different PV generations are converted to a common uniform-rank domain where it is convenient to investigate their dependence structure. It can be observed that the correlation coefficient between PV_1 and PV_2 is 0.7041, which is the strongest, so the PV arrays PV_1 and PV_2 are coupled using a Gaussian Copula function.

The superiority of Gaussian Copula can be visualized in Figure 8.2 which portrays the CDF and PDF of PV_1 and PV_2. For the Gaussian Copula, there exists a "general formula" for any number of variables. The use of Gaussian Copula function is somehow superior and best fits the adjacent PV generation data. Therefore, the Gaussian Copula function has been chosen due to its usefulness and easy implementation.

Contrary to PV_1 and PV_2, the PMCC between PV_1 and PV_3 is lower. Therefore, the dependence structure can be modeled using a Gumbel Copula exhibiting stronger tail dependence as shown in Figure 8.3, which shows the CDF and PDF of PV_1 and PV_3.

TABLE 8.2
PMCC Matrix of PV Generation Data

PV Array	PV_1	PV_2	PV_3
PV_1	1	0.7041	0.5438
PV_2	0.7041	1	0.5769
PV_3	0.5438	0.5769	1

Dependence Modeling

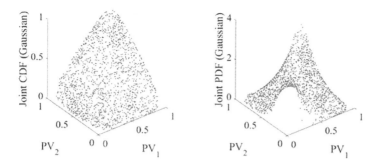

FIGURE 8.2 Joint CDF and PDF plots between PV_1 and PV_2 using Gaussian Copula.

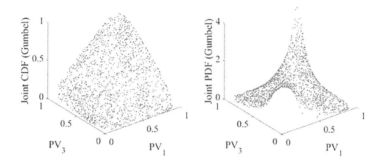

FIGURE 8.3 Joint CDF and PDF plots between PV_1 and PV_3 using Gumbel Copula.

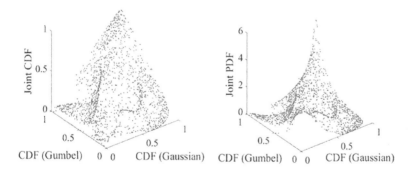

FIGURE 8.4 Joint CDF and PDF plots of Gaussian and Gumbel Copula.

Gumbel Copula, however, has an ability to capture the strong upper tail dependence and weak lower tail dependence and easily the case of independence and positive dependence; thus, it is quite good to fit in the data as a Copula function. Finally, the resulting distribution function of the coupled Gaussian and Gumbel Copula changes to Gaussian Copula as well, revealing the importance of tail shape in conditional situations as shown in Figure 8.4. It is worth noting that the mixture of diverse Copula families greatly improves the modeling feasibility, which is the attractive merit of the Copula pairing method.

8.4 CONCLUSION

A clear idea is provided about the role of stochastic dependence for PPSA using PLF. Further, different families of Copula are elaborated. Families of Copula highlighting its importance make it easier in selecting suitable Copula functions. A multivariate dependence structure is modeled using an appropriate Copula based on the algorithms that allow the inference of parameters of pair-Copula on various levels of construction. Through the detailed result analysis, the joint CDF and PDF of two different Copula functions are explained.

REFERENCES

Aas, K., Czad, C., Frigessi, A., and Bakken, H. 2009 "Pair-copula constructions of multiple dependence," *Insurance: Mathematics and Economics,* 44: 182–198.

Aas, K. and Berg, D. 2013 "Models for construction of multivariate dependence–a comparison study," in *Copulae and Multivariate Probability Distributions in Finance*: 43–64.

Fan, M., Vittal, V., Heydt, G.T., and Ayyanar, R. 2012 "Probabilistic power flow studies for transmission systems with photovoltaic generation using cumulants," *IEEE Transactions on Power Systems,* 27(4): 2251–2261.

Gupta, N., Pant, V., and Das, B. 2014 "Probabilistic load flow incorporating generator reactive power limit violations with spline based reconstruction method," *Electric Power Systems Research*, 106: 203–213.

Hourly PV generation. [Online]. Available: https://www.pvoutput.org

Jain, N., Suman, S., and Prusty, B.R. 2021 "Performance comparison of two statistical parametric methods for outlier detection and correction," *IFAC-PapersOnLine*, 54(16): 168–174.

Lu, Q., Hu, W., Min, Y., Yuan, F., and Gao, Z. 2014 "Wind power uncertainty modeling considering spatial dependence based on pair-copula theory," *In 2014 IEEE PES General Meeting Conference & Exposition*, National Harbor, MD: 1–5.

Papaefthymiou, G. and Kurowicka, D. 2009 "Using Copulas for modeling stochastic dependence in power system uncertainty analysis", *IEEE Transactions on Power Systems*, 24(1):40–49.

Prusty, B.R. and Jena, D. 2016 "Combined cumulant and Gaussian mixture approximation for correlated probabilistic load flow studies: a new approach," *CSEE Journal of Power and Energy Systems*, 2(2): 71–78.

Prusty, B.R., and Jena, D. 2017 "A critical review on probabilistic load flow studies in uncertainty constrained power systems with photovoltaic generation and a new approach," *Renewable and Sustainable Energy Reviews*, 69: 1286–1302.

Prusty, B.R. and Jena, D. 2018 "An over-limit risk assessment of PV integrated power system using probabilistic load flow based on multi-time instant uncertainty modeling," *Renewable Energy*, 116: 367–383.

Prusty, B.R. and Jena, D. 2019 "A spatiotemporal probabilistic model-based temperature-augmented probabilistic load flow considering PV generations," *International Transactions on Electrical Energy Systems*, 29(5): e2819.

Ranjan, K.G., Tripathy, D.S., Prusty, B.R., and Jena, D. 2021 "An improved sliding window prediction-based outlier detection and correction for volatile time-series," *International Journal of Numerical Modelling: Electronic Networks, Devices and Fields*, 34(1): e2816.

Dependence Modeling

Usaola, J. 2010 "Probabilistic load flow with correlated wind power injections," *Electric Power Systems Research*, 80(5): 528–536.

Wang, S., Zhang, X., and Liu, L. 2016 "Multiple stochastic correlations modeling for microgrid reliability and economic evaluation using pair-copula function." *International Journal of Electrical Power & Energy Systems* 76: 44–52.

Wu, W., et al. 2015 "A versatile probability model of photovoltaic generation using pair Copula construction," *IEEE Transactions on Sustainable Energy*, 6(4): 1337–1345.

9 Probabilistic Steady-State Analysis of Power Systems Integrated with Renewable Generations

Vikas Singh, Tukaram Moger, and Debashisha Jena
National Institute of Technology Karnataka

CONTENTS

9.1 Introduction 200
9.2 State-of-the-Art in Probabilistic Load Flow...200
    9.2.1 Review of Uncertainty Modeling in Power Systems......................200
    9.2.2 Node Power Dependency Modeling in PLF 201
9.3 Power System Models for PLF Implementation 202
9.4 Basic PLF Methods for Uncertainty Assessment 203
    9.4.1 Monte-Carlo Simulation 203
        9.4.1.1 Implementation Algorithm 204
    9.4.2 Cumulant Method..205
        9.4.2.1 Methodology for Cumulant Method 205
        9.4.2.2 Implementation Algorithm 207
    9.4.3 Point Estimation Method 209
        9.4.3.1 Location and Weight Calculations in $2n$ and $2n+1$ PEM..... 210
        9.4.3.2 Implementation Algorithm 211
9.5 Case Studies...213
    9.5.1 Sample 10-Bus Equivalent System 213
        9.5.1.1 Step-by-Step Demonstration of PEM 214
        9.5.1.2 Step-by-Step Demonstration of CM 219
        9.5.1.3 Step-by-Step Demonstration of the MCS Method............224
        9.5.1.4 Performance Evaluation, Results, and Discussion............225
    9.5.2 SR 24-Bus Equivalent system...227
9.6 Applications of Probabilistic Load Flow...232
9.7 Future Research Directions 233
9.8 Conclusions..234
Appendix A 234
References..235

DOI: 10.1201/9781003271857-9

9.1 INTRODUCTION

The present-day electrical power system is faced with different uncertain parameters (Rezaee Jordehi, 2018). The redesign of power systems, along with the integration of high-penetration renewable sources, brings greater uncertainties and incorporates new uncertain parameters in power system studies. The power system operators face issues in the decision-making process with a substantial amount of uncertain information. Thus, consideration of the uncertainties is of utmost importance for the future expansion and planning of power systems. Generally, the primary sources of power system uncertainties include outage of any conventional generating unit, change of network configurations, forecasting errors, variations of load demands, wind power, and solar photovoltaic (PV) generation.

Traditionally, the deterministic load flow (DLF) is the most fundamental tool for planning and operation of power systems, but it only considers the specific network configurations to calculate state variables and power flows. Therefore, DLF completely ignores the uncertainties in the power systems and fails to represent the actual status of a power network, which leads to an impractical estimation of the system's performance.

To perform realistic modeling for making better decisions in a power system, the probabilistic load flow (PLF) approach provides the opportunity to include the uncertainties and gives an accurate description of the functional state of the power system (Borkowska, 1974). PLF measures the uncertainties in the output variables based on the uncertainties in the input variables. Uncertainty modeling, power system model creation, and the implementation of an uncertainty handling method are the three necessary requirements for performing PLF analysis (Prusty and Jena, 2017a). With a growing interest in modeling the uncertainties in power systems, the PLF has acquired comprehensive applications in numerous areas of power systems. The significant applications of PLF include power system operation and expansion planning, three-phase PLF studies, electric railway systems, voltage and reactive power control devices, reliability evaluation, etc.

9.2 STATE-OF-THE-ART IN PROBABILISTIC LOAD FLOW

In this section, a summary of the previous work carried out in PLF analysis is presented. The PLF research performed prior to 1988 focused largely on concerns such as linearization of load flow equations, interdependence between the inputs, and the effects of the network outages. However, since 1989, numerous studies in literature have shifted their prime attention towards efficiency and accuracy of the PLF methods. The following subsections provide a detailed description of the PLF requirements and its implementations for the uncertainty assessment in power system studies.

9.2.1 Review of Uncertainty Modeling in Power Systems

The uncertainties in modern power systems are mainly classified as node power uncertainty and network uncertainty (Prusty and Jena, 2017a). Node power uncertainties occur due to outage of conventional generating units and variation of system

Probabilistic Steady-State Analysis   **201**

load demands. However, network uncertainties mainly ascend from the outage of any network element or change in transmission branch parameters due to environmental factors. The uncertainties in the conventional generation system usually follow discrete distribution and are modeled using capacity outage probability table (COPT). On outages, COPT arranges the capacity states of generating units in increasing order, along with their likelihood of occurring (Wang and McDonald, 1994). However, the randomness of the renewable energy sources is represented using continuous distribution. Weibull distribution and beta distribution are used to model the uncertainties in wind speed and solar PV generation, respectively. The uncertainty in load stems primarily from measurement and forecasting errors. When historical data is unavailable, a Gaussian distribution with the expected values of loads chosen deterministically and standard deviation equal to ±5% to ±10% of the expected value is a common assumption for load modeling (Aien and Firuz-Abad, 2015). The majority of PLF approaches disregard the consequences of network outages, thereby ignoring the risk of losing network elements such as transmission lines, transformers, and so on (Leite da Silva et al., 1985). The uncertainties associated with the network outages are often modeled using discrete distribution. The most challenging part in modeling the network outages is selecting all credible outages (Wood et al., 2013). Generally, the outage of each line or generating unit does not cause flows or voltage limit violations. The concept of contingency sequencing based on the severity order of transmission lines should be employed to avoid random checking of each line for its outage, thus reducing the amount of calculations required (Dong et al., 2010a). In most PLF studies, the variations in transmission branch parameters due to environmental factors are assumed constant. In 1993, the parameter uncertainty was considered first time in the PLF study (Patra and Misra, 1993). The uniform and binary distributions are used to model the uncertainties in series and shunt parameters of the transmission line, respectively (Su, 2005). Considering the effects of uncertain ambient temperature in PLF, the output variables are precisely calculated (Prusty and Jena, 2017b).

9.2.2 Node Power Dependency Modeling in PLF

Considering the dependability of nodal powers is of paramount importance for PLF studies. Without this, PLF review for uncertainty assessment is deemed inadequate and unreliable (Allan and Al-Shakarchi, 1977b). To avoid the difficulties that exist in actual modeling, most PLF studies presume that node power dependence is linear. In reality, however, node power dependence is rarely linear. Hence, its ignorance can lead to significant errors in the PLF analysis. Various modeling techniques are available for correlated input random variables based on the linearity or non-linearity of node power dependence. To model the linear dependence among Gaussian or non-Gaussian input variables, polynomial normal transformation (PNT) and Nataf transformation are appropriate. Copula theory, on the other hand, is often used to model non-linear dependence. The Nataf transformation transforms a vector of correlated variables into a corresponding vector of standard correlated variables. After that, random numbers are extracted from these standard normal distributions and then converted back to their original independent space (Eie, 2018). It needs a correlation matrix and marginal CDFs of input variables to create its distribution model. The PNT technique sums various orders of standard random variables to represent the input random variable. It only needs statistical moments and a correlation matrix of input random variables to establish its distribution model (Cai et al., 2015). It can be used when the marginal CDF of the input variable is unavailable or difficult to obtain. To model the dependence among uncertain input variables, Copula functions first transform these variables to a uniform/common domain employing CDF transformation. After modeling the dependence in this common domain, the input random variables can be transformed back to their original space using inverse-CDF transformation (Papaefthymiou and Kurowicka, 2008). Obtaining marginal CDFs and choosing an appropriate Copula function are two steps in the Copula theory for modeling the PDF/CDF of correlated uncertain input variables. In the literature, there exist several families of Copula, such as the elliptical family (consists t-Copula, Gaussian Copula, etc.), Farlie-Gumbel-Morgenstern (FGM) family, and the Archimedean family (consists Clayton Copula, Frank Copula, etc.). The Gaussian Copula is frequently used Copula because of its simplicity. The ability of the Student's t-Copula to capture tail dependency made it superior to the Gaussian Copula. Archimedean Copulas have a wide variety of applications due to their basic form, ease of design, flexibility in capturing different dependency structures, and many other favorable properties. To get more information on the properties of various Copula families, one can refer to the study by Nelsen (2007), Balakrishnan and Lai (2009), and Gui (2009). Choosing the right Copula to best fit the data is a significant task. Many parametric and non-parametric approaches for testing Copula fitness are available in the literature (Choroś et al., 2010).

9.3 POWER SYSTEM MODELS FOR PLF IMPLEMENTATION

To obtain better computational performance, PLF employs either a non-linear algebraic equation-based model or a simplified version of the actual model. The modeling formulations vary depending on the type of the system, such as transmission or distribution. The first implementation of PLF was performed on a DC model in 1974 (Borkowska, 1974; Allan et al., 1974). The DC models have been widely used for several years to achieve fast but approximate results. The major limitation of using the DC model is that only real power-related variables can be calculated; it does not give any information about the voltage magnitudes or reactive power flows. The DC formulations have been extended to the AC model to evaluate state variables and power flows, but the linearization process used in the problem results in a significant error at the output (Allan and Al-Shakarchi, 1976). The formulation proposed by Allan and Al-Shakarchi (1977a) has precisely calculated the statistical properties of output random variables but takes too much computation time. A fast Fourier transform-based linearization technique yields reliable and accurate results with less computational effort (Allan et al., 1981). In linear models, the selection of linearization point affects the accuracy of the PLF solution. For a large variation in input data, the point of linearization shifts towards the tail side of the distribution. In such situations, linear models performing the linearization around the mean value becomes less accurate in transforming the input data (Yuan et al., 2011). The PLF performed with

Probabilistic Steady-State Analysis

multiple points for linearization or using quadratic approaches can be used to mitigate the errors introduced by the basic linear models. A linear model based on multiple points of linearization can account for the non-linear effects of power flow equations and is employed to obtain the probability density function (PDF) of the result variables (Allan and Leite da Silva, 1981). The range limits of result variables have been accurately computed using the boundary load flow (BLF) algorithm. In contrast, quadratic PLF (QPLF) improves the accuracy by including the second-order term of the Taylor series extension of the power flow equations (Brucoli et al., 1985; Chen et al., 2008). A Gauss-QPLF is employed to handle the complexities and correlations in load demands, renewable generations, and electric vehicle charging (Gupta, 2018).

9.4 BASIC PLF METHODS FOR UNCERTAINTY ASSESSMENT

Probabilistic methods are the oldest and most popular methods for dealing with uncertain parameters in power systems. PLF was first introduced in 1974 with the intention of establishing a stronger basis for decision-making in planning problems (Borkowska, 1974). Later, its applications have been explored to power system operational and expansion planning, reliability evaluation and various other areas (Chen et al., 2008). These methods used certain PDFs to characterize the uncertainty associated with input random variables. Depending on whether to solve the problem numerically, analytically or approximately, the PLF methods are classified into three categories:

1. Numerical and sampling methods: They consist of Monte-Carlo simulation (MCS), Latin hypercube sampling (LHS)-based MCS, uniform design sampling (UDS)-based MCS, quasi-MCS (QMCS), Latin supercube sampling (LSS)-based MCS, etc.
2. Analytical methods: They consist of cumulant method (CM), convolution method, etc.
3. Approximate methods: They consist of unscented transformation method (UTM), point estimation method (PEM), etc.

In this section, the most commonly used PLF approaches like MCS, PEM, and CM are illustrated in detail along with their implementation procedures.

9.4.1 MONTE-CARLO SIMULATION

MCS is a numerical approach based on statistics and probability theory and is used for solving complex and non-linear problems (Li et al., 2013). This method evaluates the PLF indices by simulating the random behavior and the actual procedure of the system. The method, thus, addresses the problem as a set of experiments. Simply speaking, this method uses a set of random numbers to iteratively evaluate a DLF model without simplifying the original non-linear load flow equations (Dong et al., 2010b). The key benefits of using MCS method for simulation are as follows:

- MCS with simple random sampling provides the most accurate results and therefore is used as a reference for validating and comparing other PLF methods.

- The number of samples required to obtain the accurate values is independent of the size of the system which makes this method suitable for large-scale power systems.
- Non-electrical system variables like reservoir working conditions, environmental effects, and so on can be simulated as well.

Despite its many advantages, the MCS method has a major drawback of huge computation time for performing the PLF that makes this method unsuitable for use in real-world systems.

9.4.1.1 Implementation Algorithm

The step-by-step implementation steps for performing the PLF using MCS method is presented here. For n_s number of required samples, the flow chart of the algorithm is illustrated in Figure 9.1.

1. Assign a suitable PDF to each uncertain input variable to model its uncertainty (e.g. assign normal distribution to load).
2. For each input variable, perform the sampling and extract appropriate number of samples from the corresponding distribution.

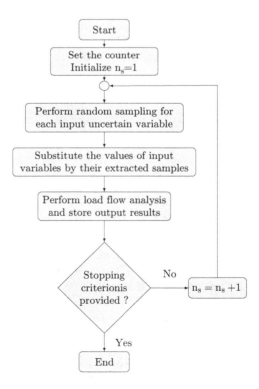

FIGURE 9.1 The computational flow chart of the Monte-Carlo method.

Probabilistic Steady-State Analysis

3. Modify the deterministic value of input variable by a corresponding set of extracted samples and then perform the load flow analysis.
4. Adjust the set of output samples into the same distribution as that of input and then find out the statistical properties of output variables.

9.4.2 CUMULANT METHOD

A widely renowned analytical approach, namely the cumulant method, requires only one DLF run to compute the cumulants of associated input uncertain variables (Sanabria and Dillon, 1986). This fact signifies the high computational efficiency of the CM for both small and large power systems. In CM, the load flow equations are linearized around the mean values of input and state variables. When there is a large variations in inputs, the dominating point of linearization shifts far away from the expected value, and the method shows poor performance in obtaining the accurate output PDFs (Dopazo et al., 1975). As the integration of renewable energy generations is increasing drastically, numerous research studies have been undertaken to resolve this issue. An enhanced CM is suggested where the input variables with operating points far from the mean values are handled separately by decomposing their components (Oke and Thomas, 2012). Using a K-means clustering approach, probabilistic optimal power flow (POPF) problems with large variations in dependent input uncertain variables can be solved (Deng et al., 2017). Bhat et al. (2017) incorporate multiple correlation among input random variables and applied CM for PLF assessment in distribution and transmission systems. The added advantage of the CM is that after acquiring the cumulants at the output, various types of series expansion methods such as Cornish-Fisher expansion (CFE), Edgeworth series expansion (ESE), Gram-Charlier expansion (GCE), etc. can be utilized to approximate the PDFs/CDFs of the result variables (Prusty and Jena, 2017a). The major drawback of the GCE is that it can only approximate unimodal distributions. However, other approximation techniques such as Laplace transform (LT) (Kenari et al., 2017), Gaussian mixture approximation (GMA) (Prusty and Jena, 2016), etc. can approximate both unimodal and multimodal distributions.

9.4.2.1 Methodology for Cumulant Method

It is well known that load flow equations are non-linear in nature, and the Newton-Raphson (NR) is a frequently used iterative process for solving these set of equations (Singh et al., 2018). To achieve a direct relation between state variables (X) and power flows (Z) with the node power inputs (W), linearization of node power equations and line power flow equations is required. The node power equations for bus p and line power flow equations between buses p and q are represented using (9.1) and (9.2), respectively.

$$P_p = \sum_{q=1}^{n_b} |V_p||V_q||Y_{pq}|\cos(\theta_{pq} - \delta_p + \delta_q)$$

$$Q_p = -\sum_{q=1}^{n_b} |V_p||V_q||Y_{pq}|\sin(\theta_{pq} - \delta_p + \delta_q)$$

$$(9.1)$$

$$P_{pq} = V_p V_q \left(G_{pq} \cos \theta_{pq} + B_{pq} \sin \theta_{pq} \right) - t_{pq} G_{pq} B_{pq} V_p^2$$

$$Q_{pq} = V_p V_q \left(G_{pq} \sin \theta_{pq} - B_{pq} \cos \theta_{pq} \right) + \left(t_{pq} B_{pq} - b_{pq0} \right) V_p^2 \tag{9.2}$$

where n_b indicates the total number of buses in the system; $|V_p|$ and $|V_q|$ denote the bus voltage magnitudes at buses p and q, respectively; Y_{pq} represents the elements of the bus admittance matrix, G represents conductance, and B represents the susceptance of the transmission lines; and b_0 and t indicate the ground susceptance and off-nominal transformer tap values, respectively.

Linearizing the set of equations in (9.1) around the mean values of node power injection (W_0) gives

$$\Delta X = S_0 \Delta W \tag{9.3}$$

where S_0 is the sensitivity matrix and can be find out from the inverse of the Jacobian matrix as $S_0 = J_0^{-1}$.

Now to get an equivalent relationship for the power flows as same as computed for state variables in (9.3), linearizing the set of equations in (9.2) around the expected values of $X(X_0)$ gives

$$\Delta Z = D_0 \Delta X \tag{9.4}$$

where D_0 represents a sensitivity matrix and is calculated at mean values of X as

$$D_0 = \frac{\partial Z}{\partial X}|_{X=X_0} \tag{9.5}$$

To get a direct relation between the node power inputs and transmission line flows, substituting ΔX value from Eq. (9.3) into Eq. (9.4) yields

$$\Delta Z = D_0 S_0 \Delta W = T_0 \Delta W \tag{9.6}$$

Thus, the set of Eqs. (9.3) and (9.6) maintains the linear relation between the output variables (state variables and line power flows) and input variables (node power inputs) and builds the basis for the CM. The CM makes use of the following properties of cumulants to describe the randomness associated with the output variables.

Property 1

The ith-order cumulant of a random variable x which is a sum of two independent random variables x_1 and x_2 with their ith-order cumulants given by $C_{x_1}^i$ and $C_{x_2}^i$, respectively, is

$$C_x^i = C_{x_1}^i + C_{x_2}^i \tag{9.7}$$

Probabilistic Steady-State Analysis **207**

Using the above property, the ith-order cumulant of the random variation of node power injection (ΔW) which is mainly due to generating unit outage (ΔW_g) and variations of load demands (ΔW_l) is given by

$$\Delta W^i = \Delta W_g{}^i + \Delta W_l{}^i \tag{9.8}$$

where ΔW^i contains the ith-order cumulants of all real and reactive power injections.

Property 2

Let $y = ax + b$ where x is a random variable with its ith-order cumulant x^i. Then the ith-order cumulant of y is

$$C_y^i = a^i C_x^i \text{ for } i \geq 2 \tag{9.9}$$

Using above property, the ith-order cumulant of the output random variables in Eqs. (9.3) and (9.6) are represented as

$$\Delta X^{(i)} = S_0^i \Delta W^{(i)} \tag{9.10}$$

$$\Delta Z^{(i)} = T_0^i \Delta W^{(i)} \tag{9.11}$$

where S_0^i and T_0^i can be found out from their corresponding elements such as

$$S_0^i (p, q) = \left(S_0 (p, q) \right)^i \tag{9.12}$$

$$T_0^i (p, q) = \left(T_0 (p, q) \right)^i \tag{9.13}$$

The raw moments can be used to compute the ith-order cumulants of any random variable. The calculation for cumulants in the first four orders are shown here:

$$
\begin{aligned}
C_1 &= \mu_1 \\
C_2 &= \mu_2 - \mu_1^2 \\
C_3 &= \mu_3 - 3\mu_1\mu_2 + 2\mu_1^3 \\
C_4 &= \mu_4 - 4\mu_1\mu_3 - 3\mu_2^2 + 12\mu_1^2\mu_2 - 6\mu_1^4
\end{aligned} \tag{9.14}
$$

where μ_i denotes the ith-order raw moment of a random variable. The statistical measures such as mean and standard deviation are equal to first-order cumulant and square root of second-order cumulant, respectively.

9.4.2.2 Implementation Algorithm

The step-by-step implementation steps for performing the PLF using CM method is presented here. The flow chart of the procedure is illustrated in Figure 9.2.

1. Assign a suitable PDF to each uncertain input variable to model its uncertainty (e.g. assign normal distribution to load).
2. For each input variable, perform the sampling and extract appropriate number of samples from the corresponding distribution.
3. Find out the ith-order raw moments for all set of samples. Then the cumulants can be calculated easily from raw moments using Eq. (9.14).
4. Compute the ith-order cumulants of the input (node power injections) to the system following Eq. (9.8).
5. Carry out a DLF with all input variables set to their expected values to find out the mean values of output variables (X_0, Z_0). Also, calculate the sensitivity matrix S_0 by inverting the Jacobian matrix J_0 obtained at the last iteration.
6. Compute the sensitivity matrix D_0 at expected value of X using Eq. (9.5). Once D_0 is find out, T_0 can be calculated as $T_0 = D_0 S_0$ from Eq. (9.6).
7. Compute the ith-order cumulants of the state variables and line flows using Eqs. (9.10) and (9.11), except the first order. The first-order cumulants are equal to their respective mean values and are already calculated in step 5.
8. Make use of an extension series such as GCE to approximate the PDFs/CDFs of the result variables.

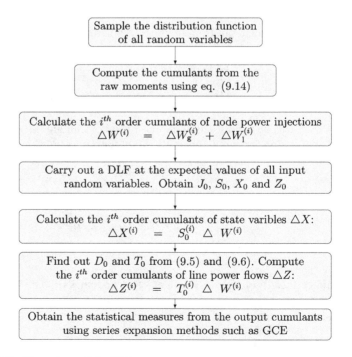

FIGURE 9.2 Flow chart of the CM.

Probabilistic Steady-State Analysis **209**

9.4.3 POINT ESTIMATION METHOD

A well-known approximation approach, namely the point estimation method, makes use of statistical moments of input uncertain variables to calculate the statistical moments of output variables. Using the PEM, the PDF of input random variables is substituted by few points, called probability concentrations. The system model uses such concentrations as input, and DLF at only those locations provides the entire picture of the randomness associated with the output variables (H.P. Hong, 1998). Rosenblueth (1975) first suggested the PEM in 1975 for dealing with symmetric random variables, and later he expanded the method to include both asymmetric and correlated random variables in 1981 (Rosenblueth, 1981). In 1989, M.E. Harr introduced another PEM which takes less number of probability concentrations than the Rosenblueth PEM (E. Harr, 1989). Harr's approach has one big flaw: it only takes symmetrical input random variables into account. In 1998, Hong also modified the Rosenblueth PEM by using a lower number of probability concentrations (H.P. Hong, 1998). The PEM gives accurate results and takes less number of DLF runs than the MCS method. Thus, the PEM achieves a reasonable compromise between computing efficiency and accuracy for the uncertainty assessment in power system studies. Different PEM methods have been proposed in the literature based on the number of points evaluated for each input uncertain variable. This chapter is mainly focused on $2n$ and $2n+1$ PEM for n number of input uncertain variables.

Considering n number of input uncertain variables $(y_1, y_2,..., y_n)$ in the system, the probabilistic output X is defined as $X = G(y_1, y_2,..., y_n)$ where G is the function expressing the relation between input and output variables. Following the theoretical background of the PEM, performing DLF runs in L number of probability concentrations for each of the input uncertain variable approximate the output variable X. For a input random variable y_j, each probability concentration l of total probability concentrations L is associated with its location and the weight. The lth location of the jth input random variable is calculated as

$$y_{j,l} = \mu_j + \xi_{j,l}\sigma_j \tag{9.15}$$

where μ_j and σ_j denote the mean and standard deviation of y_j, respectively. $\xi_{j,l}$ represents the standard location which needs to be calculated. To obtain the standard location and the weight for each probability concentration, the following set of nonlinear equations need to be solved:

$$\sum_{l=1}^{L} w_{j,l} = \frac{1}{n}$$

$$\sum_{l=1}^{L} w_{j,l}\xi_{j,l}^i = \lambda_{j,i} (i = 1, 2,..., 2L-1) \tag{9.16}$$

where, $\lambda_{j,i}$ represents the ith-order standard central moment and is calculated from the ith-order central moment (M) and standard deviation (σ) of input random variable (y_j) as

$$\lambda_{j,i} = \frac{M_{j,i}}{\sigma_j^i} \tag{9.17}$$

Note that the third- and fourth-order standard central moments represent the skewness $(\lambda_{j,3})$ and kurtosis $(\lambda_{j,4})$, respectively, for an input random variable.

After obtaining the locations and weights for all probability concentrations, DLF is run at all points $(\mu_{y_1}, \mu_{y_2}, \ldots, y_{j,l}, \ldots, \mu_{y_n})$ to yield the output random variables $(X_{j,l})$. Finally, the ith-order raw moments of $X(r_{X,i})$ can be obtained as the weighted sum of DLF results:

$$r_{X,i} = E[X^i] = \sum_{j=1}^{n} \sum_{l=1}^{L} w_{j,l} (X_{j,l})^i \tag{9.18}$$

9.4.3.1 Location and Weight Calculations in $2n$ and $2n+1$ PEM

a. **$2n$ point estimation method $(L = 2)$**

The $2n$ PEM is the most basic PEM used in the literature which replaces the PDF of each input random variable by 2 probability concentrations $(L = 2)$. From Eq. (9.16), it is noticed that for $L = 2$, the statistical knowledge provided by moments till third order is sufficient for the standard locations and weights calculation. The formulation is shown below:

$$\xi_{j,l} = \frac{\lambda_{j,3}}{2} + (-1)^{3-l} \sqrt{n + \left(\frac{\lambda_{j,3}}{2}\right)^2} \; l = 1,2 \tag{9.19}$$

$$w_{j,l} = \frac{(-1)^{3-l}}{n} \frac{\xi_{j,l}}{(\xi_{j,1} - \xi_{j,2})} l = 1,2 \tag{9.20}$$

Considering the standard locations into account, the locations of the points where DLF needs to be performed is calculated using Eq. (9.15).

It can be seen from Eq. (9.19) that standard locations are influenced by the values of input random variables (n). For large-scale power systems, the value of n rises, due to which the locations of the input random variables shifts from their mean values. These locations may be shifted to points where the PDF of input random variables is not properly defined, resulting in significant errors at the output. This limitation persist in almost all types of $L \times n$ PEM schemes and has been specified by Christian and Baecher (2002).

Probabilistic Steady-State Analysis

b. **$2n + 1$ point estimation method $(L = 3)$**

In this scheme, the PDF of each input random variable is substituted with 3 probability concentrations $(L = 3)$. However, the location of third concentration is kept fixed to its mean value by substituting the third standard location $\xi_{j,3}$ to zero in Eq. (9.15). Thus, solving Eq. (9.16) for $L = 3$ with $\xi_{j,3} = 0$ yields the standard locations and weights of y_j:

$$\xi_{j,l} = \frac{\lambda_{j,3}}{2} + (-1)^{3-l} \sqrt{\lambda_{j,4} - 3\left(\frac{\lambda_{j,3}^2}{4}\right)} l = 1,2 \tag{9.21}$$

$$\xi_{j,3} = 0 \tag{9.22}$$

$$w_{j,l} = \frac{(-1)^{3-l}}{\xi_{j,l}\left(\xi_{j,1} - \xi_{j,2}\right)} l = 1,2 \tag{9.23}$$

$$w_{j,3} = \frac{1}{n} - \frac{1}{\lambda_{j,4} - \lambda_{j,3}^2} \tag{9.24}$$

Taking the standard locations into account, the locations of the points where DLF needs to be performed are calculated using Eq. (9.15). Since, the third location for all variables are fixed to their mean values, it is enough to run only one DLF evaluation at this point, with corresponding weight w_0 equal to the sum of all the third location weights:

$$w_0 = \sum_{j=1}^{n} w_{j,3} = \sum_{j=1}^{n} \frac{1}{n} - \frac{1}{\lambda_{j,4} - \lambda_{j,3}^2} = 1 - \sum_{j=1}^{n} \frac{1}{\lambda_{j,4} - \lambda_{j,3}^2} \tag{9.25}$$

Therefore, it is pointed out that although three locations are used to define the PDF/CDF of each input random variable, it requires only $2n + 1$ DLF evaluation. Because $2n + 1$ PEM calculates standard locations and weights using moments up to the fourth order $(\lambda_{j,4})$, it delivers more accurate results than $2n$ PEM, even with only one more DLF run.

9.4.3.2 Implementation Algorithm

The step-by-step implementation steps for performing the PLF using the PEM method are presented here. The flow chart representing the procedure is illustrated in Figure 9.3.

1. Assign a suitable PDF to each uncertain input variable to model its uncertainty (e.g., assign normal distribution to load).
2. For each input variable, perform the sampling and extract appropriate number of samples from the corresponding distribution.
3. Calculate the necessary statistical measures such as skewness and kurtosis for the samples extracted from the distribution other than Gaussian (skewness for Gaussian distribution is 0 and kurtosis is 3).

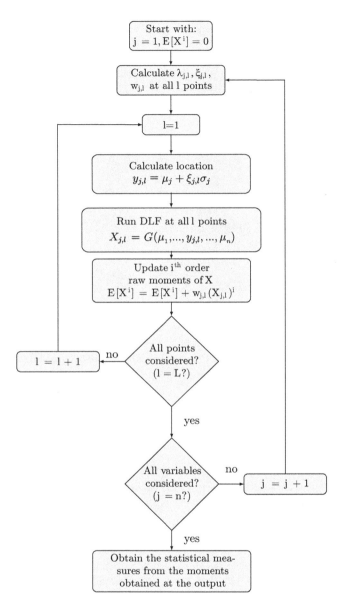

FIGURE 9.3 Flow chart of the PEM.

4. By using the information obtained from the above statistical measures, compute the standard locations, weights, and locations of the probability concentrations defined for each input random variable.
5. Run a DLF at each point (l) of all individual variable (j) with other variables fixed to their mean values, resulting in $2n$ DLF runs. Then update the ith-order raw moments of output variable (X) for each DLF run using Eq. (9.18).

6. Run one DLF with all variables fixed to their mean values. Replace $w_{j,l}$ with w_0 and update the ith-order raw moments of output variable (X) using Eq. (9.18).
7. Make use of an extension series such as GCE to approximate the PDFs/CDFs of the result variables.

9.5 CASE STUDIES

In this chapter, the case studies are performed with two separate test systems: a sample 10-bus equivalent system and a 24-bus wind integrated equivalent network of Indian southern zonal power grid. The codes for all the case studies are written in MATLAB version 2020b, with MATPOWER 6.0 (Zimmerman and Murillo-Sánchez, 2016) being used for performing the DLF evaluations. Step-by-step demonstration of the fundamental PLF techniques like MCS, PEM, and CM for a sample 10-bus equivalent system is presented.

9.5.1 SAMPLE 10-BUS EQUIVALENT SYSTEM

The single-line schematic of the system is represented in Figure 9.4. It includes seven load buses and three generator buses, with bus 1 considered as the reference bus. The base case data for the system is adopted from (Moger and Dhadbanjan, 2015).

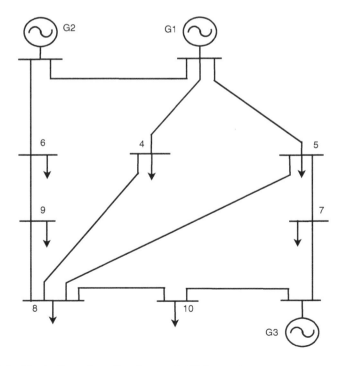

FIGURE 9.4 Single-line schematic of a sample 10-bus system.

TABLE 9.1
Base Case Data for 10-Bus System

Bus i	Type	$|V_i|$	δ_i	P_{Li}	Q_{Li}	P_{Gi}	Q_{Gi}
1	Slack	1.00	0	0	0	-	-
2	PV	1.00	0	0	0	400	-
3	PV	1.00	0	0	0	500	-
4	PQ	-	-	200	100	0	0
5	PQ	-	-	190	90	0	0
6	PQ	-	-	100	50	0	0
7	PQ	-	-	250	120	0	0
8	PQ	-	-	290	140	0	0
9	PQ	-	-	320	150	0	0
10	PQ	-	-	190	100	0	0

The probabilistic modeling of uncertain variables is represented as follows:

- Real and reactive load demands are modeled using Gaussian distribution with mean values chosen deterministically from Table 9.1 and standard deviation is considered 5% of such mean values.
- At bus 6, a wind farm with a capacity of 112.5 MW is connected. It consists of 45 wind turbines, each with a 2.5 MW capacity. Weibull distribution is employed to model the wind speed. The wind power output is considered a negative load with power factor of one. Table 9.2 displays information on wind farms.

9.5.1.1 Step-by-Step Demonstration of PEM

In this section, step-by-step implementation of the PEM ($2n$ and $2n + 1$ schemes) for a sample 10-bus equivalent system is presented.

TABLE 9.2
Wind Farm Information

Parameter		Value
Wind turbine data	P_r [MW]	2.5
	V_i [m/s]	3
	V_0 [m/s]	25
	V_r [m/s]	12.5
Wind speed data	C_W [m/s]	6.7703
	K_W	2

Probabilistic Steady-State Analysis

a. $2n+1$ Point Estimation Method

Since there are 7 load buses in the system and every real and reactive part of load is considered as a random variable, it yields 14 probabilistic inputs. Thus, according to the $2n+1$ scheme, DLF must be conducted in 29 separate locations.

Step 1: Calculate the statistical measures such as mean $\left(\mu_{y_j}\right)$, standard deviation $\left(\sigma_{y_j}\right)$, skewness $\left(\lambda_{3,y_j}\right)$ and kurtosis $\left(\lambda_{4,y_j}\right)$ of the input random variables which are necessary for the formulations of PEM. These values are summarized in Table 9.3.

Note that load demand at bus 6 is replaced with a wind farm whose mean output power is calculated using the probabilistic modeling of wind turbines shown in Appendix A, and standard deviation is assumed 52% of the mean value.

Step 2: Obtain the locations and weights of all probability concentrations for input random variables using Eqs. (9.15) and (9.21–9.25), respectively. These values are summarized in Table 9.4.

Step 3: Once the locations and weights are obtained in Table 9.4, perform the DLF at all $2n$ points following the form $\left(\mu_{y_1}, \mu_{y_2},..., y_{j,l},..., \mu_{y_n}\right)$, i.e., by keeping other variables at their mean values, resulting in $2 \times 14 = 28$ DLF runs. Because the position of the third concentration for all random variables is set to its mean value, DLF only has to be assessed once. The equivalent weight w_0 is the total of all variables' third location weights.

Step 4: Obtain the ith-order raw moments of output random variable (X) as the weighted sum of the DLF results:

TABLE 9.3
Statistical Measures for the Formulation of PEM

j	y_j	μ_{y_j}	σ_{y_j}	λ_{3,y_j}	λ_{4,y_j}
1	P_{L4}	200	0.05×200	0	3
2	P_{L5}	190	0.05×190	0	3
3	P_{L6}	55.6531	0.52×55.6531	0.6296	3.0557
4	P_{L7}	250	0.05×250	0	3
5	P_{L8}	290	0.05×290	0	3
6	P_{L9}	320	0.05×320	0	3
7	P_{L10}	190	0.05×190	0	3
8	Q_{L4}	100	0.05×100	0	3
9	Q_{L5}	90	0.05×90	0	3
10	Q_{L6}	0	0.05×0	0	3
11	Q_{L7}	120	0.05×120	0	3
12	Q_{L8}	140	0.05×140	0	3
13	Q_{L9}	150	0.05×150	0	3
14	Q_{L10}	100	0.05×100	0	3

216 Renewable Energy Integration to the Grid

TABLE 9.4

Locations and Weights of All Probability Concentrations

j	y_j	$y_{j,1}$	$w_{j,1}$	$y_{j,2}$	$w_{j,2}$
1	P_{L4}	217.3205	0.1667	182.6795	0.1667
2	P_{L5}	206.4545	0.1667	173.5455	0.1667
3	P_{L6}	-112.2774	0.1524	-17.0731	0.2237
4	P_{L7}	271.6506	0.1667	228.3494	0.1667
5	P_{L8}	315.1147	0.1667	264.8853	0.1667
6	P_{L9}	347.7128	0.1667	292.2872	0.1667
7	P_{L10}	206.4545	0.1667	173.5455	0.1667
8	Q_{L4}	108.6603	0.1667	91.3397	0.1667
9	Q_{L5}	97.7942	0.1667	82.2058	0.1667
10	Q_{L6}	0	0	0	0
11	Q_{L7}	130.3923	0.1667	109.6077	0.1667
12	Q_{L8}	152.1244	0.1667	127.8756	0.1667
13	Q_{L9}	162.9904	0.1667	137.0096	0.1667
14	Q_{L10}	108.6603	0.1667	91.3397	0.1667

$$r_{X,i} = E\left[X^i\right] = \sum_j \sum_l w_{j,l}\left(X_{j,l}\right)^i \tag{9.26}$$

For example, for voltage at bus-6 (V_6), the first- and second-order raw moments are calculated as

$$V_6^{(1)} = \sum_j \sum_l w_{j,l} V_{6_{j,l}} = 0.9779 \tag{9.27}$$

$$V_6^{(2)} = \sum_j \sum_l w_{j,l} V_{6_{j,l}}^2 = 0.9563 \tag{9.28}$$

Step 5: Obtain the cumulants of the output variables from their raw order moments using the relationship given in Eq. (9.14). The first-order cumulant (C_1) gives the mean of the output variable, and the second-order cumulant's square root $\left(\sqrt{C_2}\right)$ gives the standard deviation. Thus, the mean values and the standard deviations of all the output voltage magnitudes and voltage angles are shown in Table 9.5 and power flows in Table 9.6.

Step 6: Approximate the PDFs/CDFs of the result variables by GCE series up to the fourth order.

b. **2n Point Estimation Method**

The implementation steps for this scheme are almost similar to $2n+1$ scheme except that it uses only 2 probability concentrations for each random variable. For a given system with 14 probabilistic inputs (7 complex loads), DLF needs to be performed in $2 \times 14 = 28$ separate locations.

Probabilistic Steady-State Analysis 217

TABLE 9.5
Mean and Standard Deviation Results of the Voltage Magnitudes and Angles

Bus No.	Mean Values		Standard Deviation	
	VM	VA	VM	VA
1	1.0000	0	0	0
2	1.0000	1.0862	0	0.2610
3	1.0000	−3.8983	0	0.5008
4	0.9193	−6.9001	0.0019	0.2796
5	0.9414	−7.1253	0.0015	0.3403
6	0.9779	−2.3983	0.0013	0.5257
7	0.9621	−7.8129	0.0015	0.5192
8	0.9357	−9.4202	0.0022	0.4612
9	0.9336	−8.9429	0.0026	0.5294
10	0.9522	−8.5194	0.0017	0.5016

TABLE 9.6
Mean and Standard Deviation Results of the Active and Reactive Power Flows

From Bus	To Bus	Mean Values		Standard Deviation	
		P_{ij}	Q_{ij}	P_{ij}	Q_{ij}
1	2	−93.7528	−14.6743	22.5049	2.7065
1	4	294.9854	139.0130	11.1108	4.8246
1	5	305.3259	84.2975	13.8537	3.9544
2	6	306.0590	63.2975	22.5951	5.7988
3	7	230.0215	73.3421	5.7303	5.1105
3	10	269.9785	105.2126	5.7303	5.4685
4	8	90.0699	−86.9000	9.1687	2.8780
5	7	22.0785	−95.1414	8.6331	2.6937
5	8	88.7887	−38.0704	7.3129	3.1199
6	9	359.6695	92.4326	13.0098	4.9276
8	9	−35.0518	−8.4813	13.5155	4.9084
8	10	−76.9330	−90.8114	8.8737	4.1371

Step 1: On performing the same probabilistic modeling for loads and wind turbines as performed in $2n+1$ PEM, the mean $\left(\mu_{y_j}\right)$, standard deviation $\left(\sigma_{y_j}\right)$, and skewness $\left(\lambda_{3,y_j}\right)$ values of the input random variables are obtained and summarized in Table 9.7.

Step 2: Obtain the locations and weights for all input random variables using Eqs. (9.15) and (9.19 and 9.20), respectively. These values are described in Table 9.8.

TABLE 9.7
Statistical Measures for the Formulation of PEM

j	y_j	μ_{y_j}	σ_{y_j}	λ_{3,y_j}
1	P_{L4}	200	0.05×200	0
2	P_{L5}	190	0.05×190	0
3	P_{L6}	55.6531	0.52×55.6531	0.6296
4	P_{L7}	250	0.05×250	0
5	P_{L8}	290	0.05×290	0
6	P_{L9}	320	0.05×320	0
7	P_{L10}	190	0.05×190	0
8	Q_{L4}	100	0.05×100	0
9	Q_{L5}	90	0.05×90	0
10	Q_{L6}	0	0.05×0	0
11	Q_{L7}	120	0.05×120	0
12	Q_{L8}	140	0.05×140	0
13	Q_{L9}	150	0.05×150	0
14	Q_{L10}	100	0.05×100	0

TABLE 9.8
Locations and Weights of All Probability Concentrations

j	y_j	$y_{j,1}$	$w_{j,1}$	$y_{j,2}$	$w_{j,2}$
1	P_{L4}	237.4166	0.0357	162.5834	0.0357
2	P_{L5}	225.5457	0.0357	154.4543	0.0357
3	P_{L6}	−172.2951	0.0387	42.9446	0.0327
4	P_{L7}	296.7707	0.0357	203.2293	0.0357
5	P_{L8}	344.2540	0.0357	235.7460	0.0357
6	P_{L9}	379.8665	0.0357	260.1335	0.0357
7	P_{L10}	225.5457	0.0357	154.4543	0.0357
8	Q_{L4}	118.7083	0.0357	81.2917	0.0357
9	Q_{L5}	106.8375	0.0357	73.1625	0.0357
10	Q_{L6}	0	0.0357	0	0.0357
11	Q_{L7}	142.4499	0.0357	97.5501	0.0357
12	Q_{L8}	166.1916	0.0357	113.8084	0.0357
13	Q_{L9}	178.0624	0.0357	121.9376	0.0357
14	Q_{L10}	118.7083	0.0357	81.2917	0.0357

Probabilistic Steady-State Analysis

219

Step 3: Once the locations and weights are obtained in Table 9.8, perform the DLF at all $2n$ points following the form $\left(\mu_{y_1},\mu_{y_2},\ldots,y_{j,l},\ldots,\mu_{y_n}\right)$, i.e. by keeping other variables at their mean values, resulting in $2 \times 14 = 28$ DLF runs.

Step 4: Follow steps 4 and 5 of $2n+1$ PEM for the calculations of ith-order raw moments and cumulants of output random variables. From the first- and second-order cumulants, the expected values and standard deviations of the state variables and power flows are obtained and shown in Tables 9.9 and 9.10, respectively.

Step 5: Approximate the PDFs/CDFs of the result variables by GCE series up to the fourth order.

9.5.1.2 Step-by-Step Demonstration of CM

Step 1: Perform a DLF using the NR method with all input variables fix to their mean values to obtain the mean values of state variables (X_0) and power flows (Z_0).

$$
X = \begin{bmatrix} \delta_2 \\ \delta_3 \\ \delta_4 \\ \delta_5 \\ \delta_6 \\ \delta_7 \\ \delta_8 \\ \delta_9 \\ \delta_{10} \\ V_4 \\ V_5 \\ V_6 \\ V_7 \\ V_8 \\ V_9 \\ V_{10} \end{bmatrix},\
Z = \begin{bmatrix} P_{12} \\ P_{14} \\ P_{15} \\ P_{26} \\ P_{37} \\ P_{3,10} \\ P_{48} \\ P_{57} \\ P_{58} \\ P_{69} \\ P_{89} \\ P_{8,10} \\ Q_{12} \\ Q_{14} \\ Q_{15} \\ Q_{26} \\ Q_{37} \\ Q_{3,10} \\ Q_{48} \\ Q_{57} \\ Q_{58} \\ Q_{69} \\ Q_{89} \\ Q_{8,10} \end{bmatrix},\
X_0 = \begin{bmatrix} 1.0865 \\ -3.8968 \\ -6.8994 \\ -7.1243 \\ -2.3977 \\ -7.8114 \\ -9.4187 \\ -8.9414 \\ -8.5178 \\ 0.9193 \\ 0.9414 \\ 0.9779 \\ 0.9621 \\ 0.9357 \\ 0.9337 \\ 0.9522 \end{bmatrix},\
Z_0 = \begin{bmatrix} -93.7828 \\ 294.9689 \\ 305.3013 \\ 306.0392 \\ 230.0235 \\ 269.9765 \\ 90.0613 \\ 22.0704 \\ 88.7824 \\ 359.6616 \\ -35.0560 \\ -76.9362 \\ -14.7227 \\ 138.8963 \\ 84.1716 \\ 63.7448 \\ 73.2675 \\ 105.1073 \\ -86.9458 \\ -95.1337 \\ -38.1106 \\ 92.3468 \\ -8.5268 \\ -90.7638 \end{bmatrix}
\tag{9.29}
$$

220   Renewable Energy Integration to the Grid

TABLE 9.9
Mean and Standard Deviation Results of the Voltage Magnitudes and Angles

	Mean Values		Standard Deviation	
Bus No.	VM	VA	VM	VA
1	1.0000	0	0	0
2	1.0000	1.0968	0	0.2638
3	1.0000	−3.8907	0	0.5016
4	0.9193	−6.8956	0.0019	0.2801
5	0.9414	−7.1201	0.0015	0.3408
6	0.9779	−2.3770	0.0013	0.5312
7	0.9621	−7.8063	0.0015	0.5198
8	0.9357	−9.4110	0.0022	0.4624
9	0.9336	−8.9287	0.0026	0.5318
10	0.9522	−8.5108	0.0017	0.5026

TABLE 9.10
Mean and Standard Deviation Results of the Active and Reactive Power Flows

		Mean Values		Standard Deviation	
From Bus	To Bus	P_{ij}	Q_{ij}	P_{ij}	Q_{ij}
1	2	−94.6626	−14.5647	22.7360	2.7587
1	4	294.8065	138.9822	11.1288	4.8277
1	5	305.1142	84.2703	13.8740	3.9587
2	6	305.1455	63.8855	22.8313	5.8037
3	7	230.0755	73.3291	5.7338	5.1121
3	10	269.9245	105.1931	5.7338	5.4704
4	8	89.8956	−86.8872	9.1894	2.8797
5	7	22.0250	−95.1250	8.6350	2.6945
5	8	88.6360	−38.0621	7.3328	3.1214
6	9	360.0561	92.4802	13.0833	4.9340
8	9	−35.4287	−8.4351	13.5814	4.9159
8	10	−76.8802	−90.8056	8.8758	4.1379

where the values of voltage magnitudes (V) are obtained in per unit (p.u.) and voltage angle Δ in degrees. Equivalently, the values of active power flows (P) are obtained in MW and reactive power flows (Q) in MVAr.

Step 2: Obtain the sensitivity matrix S_0 by inverting the last iteration of Jacobian matrix J_0. The Jacobian matrix for the 10-bus equivalent system is represented using Eq. (9.30). Also, obtain the sensitivity matrix D_0 using Eq. (9.5). The size of the D_0 matrix is very big; thus, it is not shown here due to space limitations.

Step 3: Calculate the ith-order cumulants of the net power injections using Eq. (9.8). The second-order cumulants obtained for the desired buses are shown in Eq. (9.31).

Jacobian matrix for the sample 10-bus equivalent system is shown here:

$$
J_0 = \begin{bmatrix}
98.21 & 0 & 0 & 0 & -48.62 & 0 & 0 & 0 & 0 & 0 & 0 & -1.93 & 0 & 0 & 0 & 0 \\
0 & 63.47 & 0 & 0 & 0 & 31.90 & 0 & 0 & -31.58 & 0 & 0 & 0 & -1.04 & 0 & 0 & -0.63 \\
0 & 0 & 43.69 & 0 & 0 & 0 & -22.36 & 0 & 0 & 2.37 & 0 & 0 & 0 & -1.27 & 0 & 0 \\
0 & 0 & 0 & 67.15 & 0 & -22.44 & -21.87 & 0 & 0 & 0 & 4.97 & 0 & -2.05 & -1.39 & 0 & 0 \\
-48.03 & 0 & 0 & 0 & 78.31 & 0 & 0 & -30.28 & 0 & 0 & 0 & 8.64 & 0 & 0 & 0.47 & 0 \\
0 & -31.46 & 0 & -22.39 & 0 & 53.85 & 0 & 0 & 0 & 0 & -2.67 & 0 & 2.95 & 0 & 0 & 0 \\
0 & 0 & -21.18 & -21.70 & 0 & 0 & 130.12 & -43.21 & -44.03 & -3.33 & -3.24 & 0 & 0 & 10.80 & -5.01 & -5.36 \\
0 & 0 & 0 & 0 & -29.60 & 0 & -43.28 & 72.88 & 0 & 0 & 0 & -6.57 & 0 & -4.23 & 4.27 & 0 \\
0 & -31.07 & 0 & 0 & 0 & 0 & -44.17 & 0 & 75.24 & 0 & 0 & 0 & 0 & -3.97 & 0 & 5.86 \\
0 & 0 & -6.18 & 0 & 0 & 0 & 1.20 & 0 & 0 & 45.34 & 0 & 0 & 0 & -22.83 & 0 & 0 \\
0 & 0 & 0 & -8.48 & 0 & 1.97 & 1.30 & 0 & 0 & 0 & 69.42 & 0 & -23.32 & -23.37 & 0 & 0 \\
7.77 & 0 & 0 & 0 & -7.33 & 0 & 0 & -0.44 & 0 & 0 & 0 & 80.07 & 0 & 0 & -32.43 & 0 \\
0 & 5.33 & 0 & 2.51 & 0 & -7.84 & 0 & 0 & 0 & 0 & -23.78 & 0 & 53.47 & 0 & 0 & 0 \\
0 & 0 & 3.06 & 3.05 & 0 & 0 & -15.90 & 4.68 & 5.10 & -23.03 & -23.05 & 0 & 0 & 136.07 & -46.28 & -46.24 \\
0 & 0 & 0 & 0 & 6.42 & 0 & 3.96 & -10.40 & 0 & 0 & 0 & -30.26 & 0 & -46.25 & 74.84 & 0 \\
0 & 5.66 & 0 & 0 & 0 & 0 & 3.71 & 0 & -9.38 & 0 & 0 & 0 & 0 & -47.20 & 0 & 76.92
\end{bmatrix}
$$

$$(9.30)$$

The second-order cumulants of the net power injections are represented:

$$
\Delta W^{(2)} = \begin{bmatrix} 0 & 0 & 0 & 0.01 & 0.0091 & 0.0916 & 0.0151 & 0.0211 & 0.0232 & 0.0089 & 0 & 0 & 0 & 0.0025 & 0.0020 & 0 & 0.0035 & 0.0045 & 0.0053 & 0.0023 \end{bmatrix}^T
$$

$$(9.31)$$

Step 4: Compute the ith-order cumulants of the state variables (X) and power flows (Z) using Eqs. (9.10) and (9.11), respectively. The second-order cumulants of output variables are calculated and represented using Eq. (9.32).

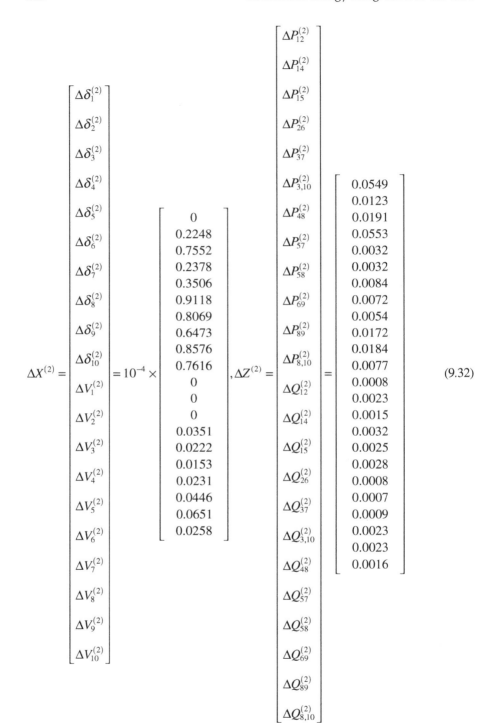

$$\Delta X^{(2)} = \begin{bmatrix} \Delta\delta_1^{(2)} \\ \Delta\delta_2^{(2)} \\ \Delta\delta_3^{(2)} \\ \Delta\delta_4^{(2)} \\ \Delta\delta_5^{(2)} \\ \Delta\delta_6^{(2)} \\ \Delta\delta_7^{(2)} \\ \Delta\delta_8^{(2)} \\ \Delta\delta_9^{(2)} \\ \Delta\delta_{10}^{(2)} \\ \Delta V_1^{(2)} \\ \Delta V_2^{(2)} \\ \Delta V_3^{(2)} \\ \Delta V_4^{(2)} \\ \Delta V_5^{(2)} \\ \Delta V_6^{(2)} \\ \Delta V_7^{(2)} \\ \Delta V_8^{(2)} \\ \Delta V_9^{(2)} \\ \Delta V_{10}^{(2)} \end{bmatrix} = 10^{-4} \times \begin{bmatrix} 0 \\ 0.2248 \\ 0.7552 \\ 0.2378 \\ 0.3506 \\ 0.9118 \\ 0.8069 \\ 0.6473 \\ 0.8576 \\ 0.7616 \\ 0 \\ 0 \\ 0 \\ 0.0351 \\ 0.0222 \\ 0.0153 \\ 0.0231 \\ 0.0446 \\ 0.0651 \\ 0.0258 \end{bmatrix}, \Delta Z^{(2)} = \begin{bmatrix} \Delta P_{12}^{(2)} \\ \Delta P_{14}^{(2)} \\ \Delta P_{15}^{(2)} \\ \Delta P_{26}^{(2)} \\ \Delta P_{37}^{(2)} \\ \Delta P_{3,10}^{(2)} \\ \Delta P_{48}^{(2)} \\ \Delta P_{57}^{(2)} \\ \Delta P_{58}^{(2)} \\ \Delta P_{69}^{(2)} \\ \Delta P_{89}^{(2)} \\ \Delta P_{8,10}^{(2)} \\ \Delta Q_{12}^{(2)} \\ \Delta Q_{14}^{(2)} \\ \Delta Q_{15}^{(2)} \\ \Delta Q_{26}^{(2)} \\ \Delta Q_{37}^{(2)} \\ \Delta Q_{3,10}^{(2)} \\ \Delta Q_{48}^{(2)} \\ \Delta Q_{57}^{(2)} \\ \Delta Q_{58}^{(2)} \\ \Delta Q_{69}^{(2)} \\ \Delta Q_{89}^{(2)} \\ \Delta Q_{8,10}^{(2)} \end{bmatrix} = \begin{bmatrix} 0.0549 \\ 0.0123 \\ 0.0191 \\ 0.0553 \\ 0.0032 \\ 0.0032 \\ 0.0084 \\ 0.0072 \\ 0.0054 \\ 0.0172 \\ 0.0184 \\ 0.0077 \\ 0.0008 \\ 0.0023 \\ 0.0015 \\ 0.0032 \\ 0.0025 \\ 0.0028 \\ 0.0008 \\ 0.0007 \\ 0.0009 \\ 0.0023 \\ 0.0023 \\ 0.0016 \end{bmatrix} \quad (9.32)$$

Probabilistic Steady-State Analysis

where the values for active power (P), reactive power Q, and voltage magnitude (V) are obtained in p.u.; voltage angle (δ) unit is in radian. The standard deviation of a random variable is calculated by taking the square root of the second-order cumulant. The p.u. values of P and Q need to be converted into conventional units of MW and MVAr, and the radian values of δ into degrees to yield the final output results. After this operation, the output values of state variables and power flows are obtained and are summarized in Tables 9.11 and 9.12.

Step 5: Approximate the PDFs/CDFs of the results variables by GCE series up to the fourth order.

TABLE 9.11
Mean and Standard Deviation Results of the Voltage Magnitudes and Angles

Bus No.	Mean Values		Standard Deviation	
	VM	VA	VM	VA
1	1.0000	0	0	0
2	1.0000	1.0865	0	0.2717
3	1.0000	−3.8968	0	0.4979
4	0.9193	−6.8994	0.0019	0.2794
5	0.9414	−7.1243	0.0015	0.3392
6	0.9779	−2.3977	0.0012	0.5471
7	0.9621	−7.8114	0.0015	0.5147
8	0.9357	−9.4187	0.0021	0.4610
9	0.9337	−8.9414	0.0026	0.5306
10	0.9522	−8.5178	0.0016	0.5000

TABLE 9.12
Mean and Standard Deviation Results of the Active and Reactive Power Flows

From Bus	To Bus	Mean Values		Standard Deviation	
		P_{ij}	Q_{ij}	P_{ij}	Q_{ij}
1	2	−93.7828	−14.7227	23.4229	2.7918
1	4	294.9689	138.8963	11.1017	4.7746
1	5	305.3013	84.1716	13.8120	3.9026
2	6	306.0392	63.7448	23.5119	5.6788
3	7	230.0235	73.2675	5.6629	5.0133
3	10	269.9765	105.1073	5.6629	5.3079
4	8	90.0613	−86.9458	9.1638	2.8186
5	7	22.0704	−95.1337	8.5059	2.6484
5	8	88.7824	−38.1106	7.3274	3.0352
6	9	359.6616	92.3468	13.1266	4.7985
8	9	−35.0560	−8.5268	13.5496	4.7781
8	10	−76.9362	−90.7638	8.7909	4.0178

9.5.1.3 Step-by-Step Demonstration of the MCS Method

In this section, the MCS method with 10,000 samples has been implemented and serves as the reference method for the performance assessment of CM and PEM in terms of accuracy and computational efficiency. MCS involves extracting a set of samples for input random variables from the corresponding probability distributions, running DLF for each set, and finally obtaining the PDF/CDF of output variables (state variables and power flows) from the results of repetitive sampling. The expected values and the standard deviations of all the output voltage magnitudes and voltage angles are shown in Table 9.13 and power flows in Table 9.14.

TABLE 9.13

Mean and Standard Deviation Results of Voltage Magnitudes and Angles

Bus No.	Mean Values		Standard Deviation	
	VM	VA	VM	VA
1	1.0000	0	0	0
2	1.0000	1.08704	0	0.266366
3	1.0000	−3.89691	0	0.503811
4	0.919261	−6.90018	0.00189243	0.278976
5	0.941349	−7.12433	0.00149546	0.34132
6	0.977907	−2.3967	0.00126993	0.536366
7	0.962086	−7.81124	0.001532	0.521227
8	0.935676	−9.41986	0.00218332	0.464558
9	0.933604	−8.94384	0.00263486	0.53657
10	0.952181	−8.51822	0.00166022	0.50536

TABLE 9.14

Mean and Standard Deviation Results of the Active and Reactive Power Flows

From Bus	To Bus	Mean Values		Standard Deviation	
		P_{ij}	Q_{ij}	P_{ij}	Q_{ij}
1	2	−93.8217	−14.664	22.9635	2.76471
1	4	294.987	139.021	11.0831	4.82684
1	5	305.286	84.3081	13.8953	3.91757
2	6	305.989	63.9554	23.056	5.84237
3	7	230.007	73.3645	5.72144	5.05427
3	10	269.993	105.222	5.72144	5.48939
4	8	90.055	−86.8823	9.26872	2.87301
5	7	22.051	−95.1334	8.66759	2.65149
5	8	88.8125	−38.0578	7.35159	3.15299
6	9	359.805	92.4413	13.0275	4.91688
8	9	−34.9592	−8.52905	13.7112	4.93975
8	10	−76.9972	−90.8369	8.95278	4.14752

Probabilistic Steady-State Analysis

9.5.1.4 Performance Evaluation, Results, and Discussion

The results of the MCS method with 10,000 samples have been used as a reference for the performance assessment of CM and PEM in terms of accuracy and computational efficiency. Absolute percentage error (APE) values have been used as accuracy indices to calculate the mistakes in the statistical characteristics of output random variables (Singh et al., 2021). Its formulation is represented as

$$\text{APE}_X = \left| \frac{(X_{\text{PLF}} - X_{\text{MCS}})}{X_{\text{MCS}}} \right| \times 100\% \tag{9.33}$$

where X_{MCS} represents the output value achieved using MCS and X_{PLF} denotes the output value achieved using other PLF technique except MCS. The APE values obtained for a sample 10-bus equivalent system are shown in Table 9.15. It is noticed that the error values incurred by the $2n+1$ PEM in its statistical outputs are least as compared to other methods. Thus, it provides the most reliable results. However, it obtains the worst case result at the standard deviation of voltage angle. Since the system size is small, the $2n$ PEM also gives good results. This is due to the fact that the number of input random variables (n) affects the probability concentrations of the $2n$ scheme. As the value of n increases, the estimation error at the output increases. The CM performs poorly as it has much larger APE values because of large fluctuations in input random variables. This poor performance of CM is due to linearization requirement of load flow equations.

The voltage magnitude CDF of bus 6 is shown in Figure 9.5, and the active power flow PDF of lines 6–9 is shown in Figure 9.6. From Figure 9.6, it is observed that the PDF of $2n+1$ PEM suits the PDF of MCS better than the PDFs of CM and $2n$ PEM. Also, it is observed from Figure 9.5 that $2n+1$ PEM gives more accurate findings till first two order cumulants, but not for the higher-order cumulants like third and fourth. This leads to substantial errors in the distribution graphs of result variables.

TABLE 9.15
Selected Percentage Error Values for a Sample 10-Bus System

μ and σ Outputs (10-Bus System)		V_6 (p.u.)	δ_6 (degrees)	P_{69} (MW)	Q_{69} (MVAr)
CM	μ	0.00071	0.0417	0.0399	0.1022
	σ	5.5066	2.0012	0.7607	2.4076
$2n+1$ PEM	μ	0.00071	0.0668	0.0377	0.0094
	σ	2.3678	1.9886	0.1359	0.2180
$2n$ PEM	μ	0.00071	0.8220	0.0698	0.0421
	σ	2.3678	0.9631	0.4283	0.3482

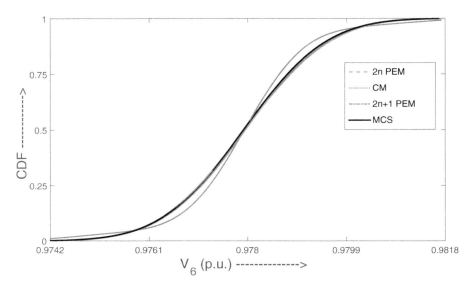

FIGURE 9.5 CDF plot of the voltage magnitude at bus 6.

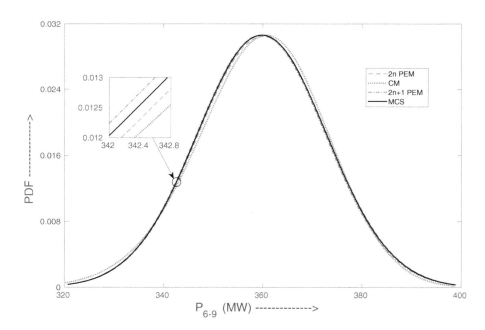

FIGURE 9.6 PDF plot of the active power flow in lines 6–9.

9.5.2 SR 24-Bus Equivalent system

A 24-bus wind integrated equivalent network of southern region of India (SR 24-bus equivalent system) has been considered to validate the implementation of the fundamental PLF approaches. The single-line diagram of the system is represented in Figure 9.7. The necessary data for the system are adopted from Moger

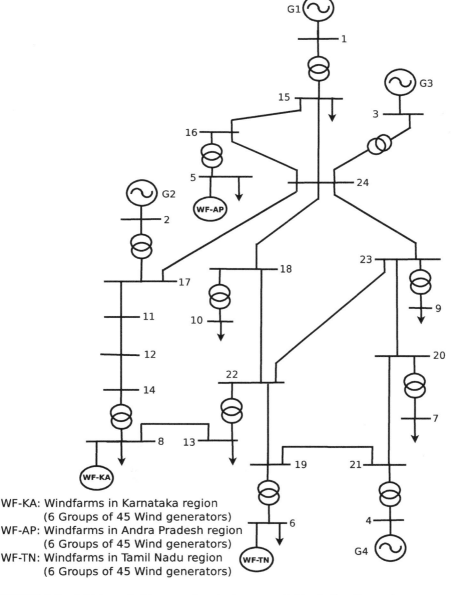

WF-KA: Windfarms in Karnataka region
(6 Groups of 45 Wind generators)
WF-AP: Windfarms in Andra Pradesh region
(6 Groups of 45 Wind generators)
WF-TN: Windfarms in Tamil Nadu region
(6 Groups of 45 Wind generators)

FIGURE 9.7 A 24-bus wind integrated equivalent network (single-line diagram).

and Dhadbanjan (2016). The system consists of 4 generators, 12 transformers, and 16 transmission lines that link the four southern region states of India such as Tamil Nadu, Andhra Pradesh, Karnataka, and Kerala. Three wind farms of Karnataka, Tamil Nadu, and Andhra Pradesh regions installed at buses 5, 6, and 8 are shown in Figures 9.8, 9.9, and 9.10, respectively. Each wind farm is divided into six categories, each with same sort of wind generators, either fixed speed or variable speed wind generators. Each farm has 45 wind turbines with a total rated capacity of 47.5 MW.

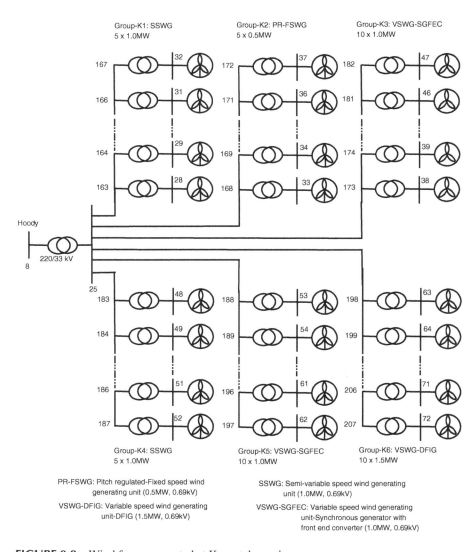

FIGURE 9.8 Wind farm connected at Karnataka region.

Probabilistic Steady-State Analysis

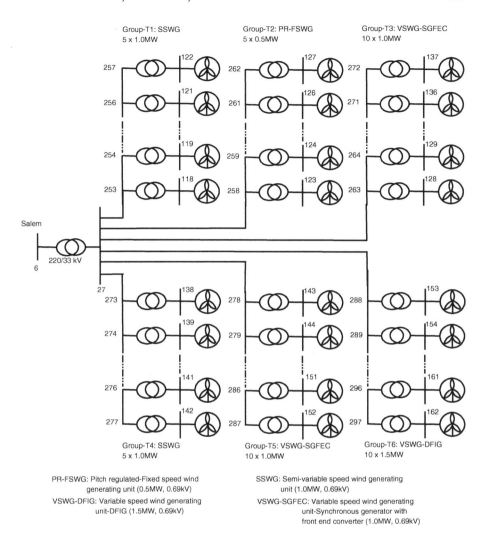

FIGURE 9.9 Wind farm connected at Tamil Nadu region.

The probabilistic modeling of the uncertain variables is as follows:

- The uncertainties related to active and reactive power load demands at all load buses are taken into consideration. The uncertainty is modeled using normal distribution where mean values are chosen deterministically, and standard deviation is considered 5% of such mean values.
- Weibull distribution is employed to model the wind speed. The wind power output is considered a negative load with a power factor of one. Table 9.2 displays information on wind farms.

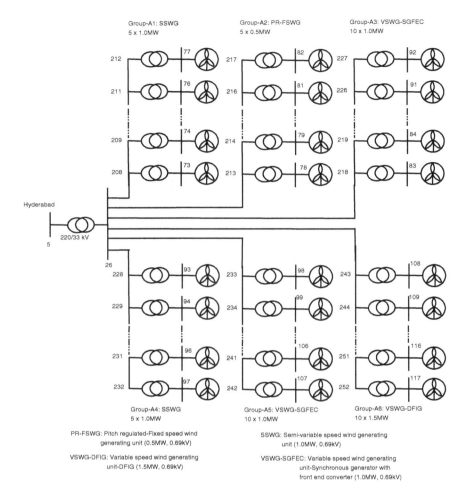

FIGURE 9.10 Wind farm connected at Andhra Pradesh region.

The mean and standard deviation values of the voltage magnitude obtained for the SR 24-bus equivalent system by the basic PLF methods are shown in Table 9.16.

The simulation time consumed by all the basic PLF methods for both the systems considered for case studies is shown in Table 9.17.

Inferences:

Table 9.16 shows that when compared to CM and $2n$ PEM, the output values of the voltage magnitude computed from $2n+1$ PEM are extremely near to those obtained by the MCS technique. As a result, the inaccuracy introduced by $2n+1$ PEM is very minimal, making this technique appropriate for large-size power systems. The voltage magnitude values obtained by CM are very far from the values obtained using MCS. This poor performance of CM when dealing with large input variation is due to linearization requirement of the load flow equations around the mean values.

Probabilistic Steady-State Analysis

TABLE 9.16

Output Values of the Voltage Magnitude (p.u.)

Bus No.	Mean Values				Standard Deviation			
	MCS	CM	2n PEM	2n+1 PEM	MCS	CM	2n PEM	2n+1 PEM
1	1.0000	1.0000	1.0000	1.0000	0	0	0	0
2	1.0000	1.0000	1.0000	1.0000	0	0	0	0
3	1.0000	1.0000	1.0000	1.0000	0	0	0	0
4	1.0000	1.0000	1.0000	1.0000	0	0	0	0
5	0.8442	0.8514	0.8507	0.8464	0.0093	0.0089	0.0093	0.0093
6	0.8712	0.8776	0.8771	0.8732	0.0063	0.0058	0.0059	0.0059
7	0.8609	0.8660	0.8655	0.8625	0.0069	0.0066	0.0067	0.0067
8	0.8470	0.8573	0.8564	0.8497	0.0103	0.0096	0.0096	0.0100
9	0.9221	0.9286	0.9280	0.9241	0.0057	0.0053	0.0054	0.0052
10	0.9165	0.9238	0.9231	0.9188	0.0063	0.0058	0.0059	0.0059
11	0.9721	0.9782	0.9777	0.9740	0.0050	0.0045	0.0046	0.0048
12	0.9576	0.9652	0.9646	0.9589	0.0065	0.0060	0.0060	0.0063
13	0.8472	0.8566	0.8557	0.8488	0.0099	0.0093	0.0094	0.0096
14	0.8792	0.8895	0.8886	0.8801	0.0096	0.0089	0.0090	0.0095
15	0.9768	0.9787	0.9785	0.9774	0.0015	0.0014	0.0015	0.0015
16	0.9131	0.9188	0.9183	0.9149	0.0057	0.0014	0.0057	0.0057
17	0.9880	0.9917	0.9914	0.9892	0.0029	0.0026	0.0027	0.0027
18	0.9268	0.9340	0.9334	0.9291	0.0062	0.0056	0.0057	0.0057
19	0.8990	0.9048	0.9043	0.9009	0.0053	0.0050	0.0051	0.0051
20	0.8954	0.9004	0.8999	0.8970	0.0055	0.0052	0.0053	0.0053
21	0.9446	0.9481	0.9478	0.9457	0.0033	0.0031	0.0031	0.0031
22	0.8811	0.8895	0.8888	0.8837	0.0082	0.0076	0.0077	0.0078
23	0.9381	0.9445	0.9439	0.9401	0.0055	0.0051	0.0051	0.0051
24	0.9683	0.9723	0.9720	0.9696	0.0031	0.0029	0.0030	0.0028

The $2n$ PEM also shows significant errors due to increased value of number of input random variables (n) for large-size network.

From Table 9.17, it is noticed that CM requires the least amount of computation time for both the test systems under consideration and is thus regarded as the most computationally efficient approach. Despite the fact that the $2n+1$ PEM takes one additional DLF run than $2n$ PEM, it takes less computation time.

TABLE 9.17

Computation Time (sec)

Computation Time	MCS	CM	2n+1 PEM	2n PEM
10-bus system	127.8290	0.060	0.2710	0.3310
24-bus system	341.1990	0.090	0.8150	0.8480

9.6 APPLICATIONS OF PROBABILISTIC LOAD FLOW

With an increasing interest in modeling the power system uncertainties, the PLF has acquired comprehensive applications in numerous areas of power systems. The main application fields of PLF methods in power systems are as follows.

Power system expansion and operational planning: Due to an increasing penetration of intermittent renewable generations in modern power systems, power system planning is becoming a crucial subject and a popular challenge. The key aim of the power system planning is to fulfill the consumer load demands at reduced cost of generation, transmission, and distribution by satisfying the technical, political, and economic constraints (Dong et al., 2010b). Since the behavior of power systems is stochastic, computational system planning should be performed using probabilistic approaches. Presently, most of the design, planning, and operation strategies in power systems are based on deterministic approaches that find difficulty in dealing with the uncertainties and addressing other power system challenges. The power system planning using probabilistic approaches offers a realistic and efficient technique by considering the stochastic nature of load demands and/or component outages. Also, PLF indices gives the probabilistic information of bus voltages violating its limit, power flow violating its thermal rating, etc. to improve the reliability of the power system.

Railway electrification system: An electrified railway system is analogous to power transmission and distribution system except that loads (trains) travel and change its mode of operation regularly. A moving train's power usage is primarily determined by its speed which is in turn linked to operating conditions and system configurations. The same load (train) becomes a source when regenerative breaking is allowed (Prusty and Jena, 2017a). Thus, power supplied and consumed by the train varies in a wide range and is completely uncertain. Other factors that lead to uncertainties in railway systems include traffic requirements, driver's behavior, track configuration, etc. In such conditions, PLF considers train position or other uncertain parameter as input random variable and solves the system that helps in service, operation, and extension of the railway system.

Reactive power and voltage control devices: The reactive power and node voltage of entire power systems can be regulated by preserving the voltage within the desired range and diminishing the losses. Reactive power and voltage control equipment such as static var compensators, switched capacitor banks, and tap-changing transformers are used in power systems. For obtaining better voltage regulation and minimized losses in the network, these control devices should be tuned based on the probabilistic approach rather than the deterministic approach. The probabilistic approaches not only find the operating constraint violations but also the likelihood of each violation (Chen et al., 2008).

Three-phase probabilistic load flow: In modern power systems, voltage unbalance has emerged as one of a variety of causes that degrade the efficiency of electric power and thus requires careful attention. Asymmetry in impedances of transmission lines and loads inequality in three phases are the main reasons for the unbalancing of three-phase voltages. In order to assess the unbalancing of voltage in the distribution systems, it is necessary to take into account the inevitable uncertainties associated with the input data. To evaluate the effect of uncertainties in an unbalanced

Probabilistic Steady-State Analysis

distribution system, the normal PLF analysis should be extended to area of three-phase under steady-state condition. In 1997, three-phase PLF was introduced for the first time and MCS was used to calculate the voltage unbalance caused by load uncertainty (Wang and Pierrat, 1997).

Power system stability analysis: Depending on the level of disturbances introduced to the power system, the stability analysis is primarily characterized as small-signal stability and transient stability (Kundur, 2007).

For a long time, the transient stability analysis for power systems was performed deterministically and has been widely used for dynamic security analysis. Many severe operational decisions such as choosing a fault type, deciding a fault location and contingency selection were selected manually based on experience (Dong et al., 2010b). While the deterministic approach has provided adequate results in the power sector, it completely overlooks the stochastic character of a real power system. In addition to varying nature of load demands, the outage of generating units and branch outages also contribute to the uncertainties which greatly affect the performance of transient stability analysis. Thus, deterministic approaches are no longer suitable for stability assessment of a sophisticated system. The transient stability using probabilistic approaches has become extremely significant for power system stability analysis.

The study of small-signal stability is critical for ensuring the safe and healthy functioning of power systems with increasing uncertainties. The small-signal stability analysis gives good performance when accurate models are used. The deterministic approaches overlook the stochastic character of power systems; thus, small-signal stability analysis using probabilistic approaches is getting more and more attention.

9.7 FUTURE RESEARCH DIRECTIONS

Following the theoretical framework and examining the case studies thoroughly, the following suggestions are made for future research directions:

- The complexities in the input data are steadily growing as a result of increasingly innovative improvements in the regulatory framework of power systems and technological innovation. Thus, exploring and modeling the new input uncertainties to obtain realistic outcomes in power system studies are proposed for future studies.
- Some important sources of uncertainty such as conventional generation and branch outages are not addressed in this chapter and left as a suggestion for future work to analyze their influence on the output results in power systems.
- In this chapter, the basic PLF methods are employed for the uncertainty assessment. Modifying these techniques or hybridizing them to obtain more accurate results by consuming less computation time is advised for future research.
- In this chapter, the PLF techniques are discussed with independent input random variables. Practically, in most of the cases, power system uncertain parameters are correlated, and this correlation is hardly linear. Modeling the actual dependence among different uncertain parameters using suitable technique is worth investigating for future work.

- Because of the power industry's complexities and ever-increasing uncertainties, new problems arise often, necessitating the development of new techniques. Thus, introducing new probabilistic techniques that are highly accurate with less computational burden should be the focus for future studies.
- Non-parametric density estimation methods such as kernel density estimation (KDE), adaptive KDE (AKDE), saddle point approximation (SPA), etc. do not predefine the parametric PDFs for input random variables. It provides greater flexibility to the user in modeling the input uncertainty by selecting a wide range of probability distribution functions and thus recommended for future research.

9.8 CONCLUSIONS

In this chapter, the basic probabilistic load flow methods like Monte-Carlo method, cumulant method, and point estimation method have been explored in detail for the uncertainty assessment in renewable energy-integrated power systems. The implementation steps of all these basic methods for a sample 10-bus equivalent system have been demonstrated. In addition, a 24-bus wind integrated equivalent network of Indian southern zonal power grid has been examined in order to evaluate the performance of the fundamental PLF approaches for actual power systems. Since MCS is the most accurate PLF method, its results have been used as a reference for the performance evaluation of CM and PEM. Results demonstrated that the APE values obtained from the CM and PEM increase as the degree of uncertainty increases. However, the phenomenon is more apparent for the CM because it is restricted by the linearization of load flow equations. The accuracy of $2n$ PEM diminishes as the size of the input random variables (n) increases; hence, it is unsuitable for large-size power systems. The formulation of $2n+1$ PEM uses higher-order moments and does not have such boundation of dependence on the value of n; thus, it provides the most accurate results. The PEM consumes fewer DLF runs than the MCS method but is computationally less efficient than CM. To summarize, a suitable tradeoff between accuracy and speed is always required when dealing with non-simulation methods like CM and PEM. The most appropriate approach for a given task is determined based on the priority between accuracy and computational efficiency, system size, and the degree of uncertainty associated with random input variables in the system.

APPENDIX A

PROBABILISTIC MODELING OF WIND TURBINES

Wind speed is primarily responsible for the generation of electric power from wind turbines. The wind speed uncertainty is usually modeled using Weibull distribution. In general, depending on the shaping parameters, Weibull distributions exist in various shapes. In this chapter, a two-parameter continuous Weibull distribution has been used. For a given scale parameter (α_w) and shape parameter (β_w), the PDF of the Weibull-distributed wind velocity (V) is given by

Probabilistic Steady-State Analysis

$$f_V(v) = \left(\frac{\alpha_w}{\beta_w}\right)\left(\frac{v}{\beta_w}\right)^{(\alpha_w-1)} \exp\left[-\left(\frac{v}{\beta_w}\right)^{\alpha_w}\right], v > 0 \tag{9.34}$$

To obtain the CDF of Weibull-distributed random variable V, integrate Eq. (9.34) to a value of v:

$$F_V(v) = \int_{-\infty}^{v} f_V(v) = 1 - e^{-\left(\frac{v}{\beta_w}\right)^{\alpha_w}} \tag{9.35}$$

Using the specified parameters α_w and β_w, the wind speed samples need to be extracted randomly from the Weibull PDF and then transformed to the generator output power (P_W) using wind speed-power curve defined as

$$P_W = \begin{cases} 0 & v \leq v_i \text{ or } v \geq v_0 \\[2mm] P_r \dfrac{v - v_i}{v_r - v_i} & v_i < v < v_r \\[2mm] P_r & v_r \leq v \leq v_0 \end{cases} \tag{9.36}$$

where v_i, v_r and v_0 represent the cut-in speed, rated speed, and cut-out speed, respectively, of the wind turbines; P_r denotes the rated power of the wind turbine.

The wind speed-power curve given by Eq. (9.36) is not a continuous function of v. The PDF of P_W can be obtained using Eq. (9.37), and it also follows Weibull distribution.

$$f(P_W) = \begin{cases} 1 - e^{\left[-\left(\frac{v_i}{\beta_w}\right)^{\alpha_w}\right]} + e^{\left[-\left(\frac{v_0}{\beta_w}\right)^{\alpha_w}\right]} & P_W = 0 \\[4mm] \dfrac{\alpha_w}{K_1\beta_w}\left(\dfrac{P_W - K_2}{K_1\beta_w}\right)^{\alpha_w-1} e^{\left[-\left(\frac{P_W - K_2}{K_1\beta_w}\right)^{\alpha_w}\right]} & 0 < P_W < P_{WR} \\[4mm] e^{\left[-\left(\frac{v_r}{\beta_w}\right)^{\alpha_w}\right]} - e^{\left[-\left(\frac{v_0}{\beta_w}\right)^{\alpha_w}\right]} & P_W = P_{WR} \end{cases} \tag{9.37}$$

where $K_1 = \dfrac{P_{WR}}{v_r - v_i}$ and $K_2 = -K_1 * v_i$; P_{WR} denotes wind turbine generator's rated power.

REFERENCES

Aien, M. and Firuz-Abad, M. (2015). Probabilistic optimal power flow in correlated hybrid wind-pv power systems: A review and a new approach. *Renewable and Sustainable Energy Reviews*, 41:1437–1446.

Allan, R. and Al-Shakarchi, M. (1977a). Probabilistic techniques in ac load-flow analysis. In *Proceedings of the Institution of Electrical Engineers*, volume 124, pages 154–160. IET.

Allan, R. N. and Al-Shakarchi, M. R. G. (1976). Probabilistic a.c. load flow. *Proceedings of the Institution of Electrical Engineers*, 123(6):531–536.

Allan, R. N. and Al-Shakarchi, M. R. G. (1977b). Linear dependence between nodal powers in probabilistic a.c. load flow. *Proceedings of the Institution of Electrical Engineers*, 124(6):529–534.

Allan, R. N., Borkowska, B., and Grigg, C. H. (1974). Probabilistic analysis of power flows. *Proceedings of the Institution of Electrical Engineers*, 121(12):1551–1556.

Allan, R. N. and Leite da Silva, A. M. (1981). Probabilistic load flow using multilinearisa- tions. *IEE Proceedings C - Generation, Transmission and Distribution*, 128(5):280–287.

Allan, R. N., Leite Da Silva, A. M., and Burchett, R. C. (1981). Evaluation methods and accuracy in probabilistic load flow solutions. *IEEE Transactions on Power Apparatus and Systems*, PAS-100(5):2539–2546.

Balakrishnan, N. and Lai, C. D. (2009). *Continuous Bivariate Distributions*. Springer Science & Business Media, Berlin.

Bhat, N. G., Prusty, B. R., and Jena, D. (2017). Cumulant-based correlated probabilistic load flow considering photovoltaic generation and electric vehicle charging demand. *Frontiers in Energy*, 11(2):184–196.

Borkowska, B. (1974). Probabilistic load flow. *IEEE Transactions on Power Apparatus and Systems*, PAS-93(3):752–759.

Brucoli, M., Torelli, F., and Napoli, R. (1985). Quadratic probabilistic load flow with linearly modelled dispatch. *International Journal of Electrical Power & Energy Systems*, 7(3):138–146.

Cai, D., Li, X., Zhou, K., Xin, J., and Cao, K. (2015). Probabilistic load flow algorithms considering correlation between input random variables: A review. In *2015 IEEE 10th Conference on Industrial Electronics and Applications (ICIEA)*, pages 1139–1144. IEEE.

Chen, P., Chen, Z., and Bak-Jensen, B. (2008). Probabilistic load flow: A review. In *2008 Third International Conference on Electric Utility Deregulation and Restructuring and Power Technologies*, pages 1586–1591.

Choroś, B., Ibragimov, R., and Permiakova, E. (2010). Copula estimation. In Jaworski, P., Durante, F., Härdle, W., and Rychlik T. (eds.) *Copula Theory and Its Applications*, pages 77–91. Springer, Berlin.

Christian, J. T. and Baecher, G. B. (2002). The point-estimate method with large numbers of variables. *International Journal for Numerical and Analytical Methods in Geomechanics*, 26(15):1515–1529.

Deng, X., He, J., and Zhang, P. (2017). A novel probabilistic optimal power flow method to handle large fluctuations of stochastic variables. *Energies*, 10:1623.

Dong, L., Cheng, W., Bao, H., and Yang, Y. (2010a). A probabilistic load flow method with consideration of random branch outages and its application. In *2010 Asia-Pacific Power and Energy Engineering Conference*, pages 1–4.

Dong, Z., Zhang, P., Ma, J., Zhao, J., Ali, M., Meng, K., and Yin, X. (2010b). *Emerging Techniques in Power System Analysis*. Springer, Heidelberg.

Dopazo, J. F., Klitin, O. A., and Sasson, A. M. (1975). Stochastic load flows. *IEEE Transactions on Power Apparatus and Systems*, 94(2):299–309.

Harr, M. E. (1989). Probabilistic estimates for multivariate analysis. *Applied Mathematical Modelling*, 13(5):313–318.

Eie, M. H. (2018). Probabilistic load flow studies: Analytical and approximate methods. Master's thesis, NTNU.

Gui, W. (2009). Adaptive series estimators for copula densities.

Gupta, N. (2018). Gauss-quadrature-based probabilistic load flow method with voltage- dependent loads including WTGS, PV, and EV charging uncertainties. *IEEE Transactions on Industry Applications*, 54(6):6485–6497.

Hong, H. P. (1998). An efficient point estimate method for probabilistic analysis. *Reliability Engineering and System Safety*, 59(3):261–267.

Kenari, M. T., Sepasian, M. S., Nazar, M. S., and Mohammadpour, H. A. (2017). Combined cumulants and Laplace transform method for probabilistic load flow analysis. *IET Generation, Transmission & Distribution*, 11(14):3548–3556.

Kundur, P. (2007). Power system stability. *Power System Stability and Control*, pages 7–1. McGraw-Hill, New York.

Leite da Silva, A. M., Allan, R. N., Soares, S. M., and Arienti, V. L. (1985). Probabilistic load flow considering network outages. *IEE Proceedings C - Generation, Transmission and Distribution*, 132(3):139–145.

Li, W. et al. (2013). *Reliability Assessment of Electric Power Systems using Monte Carlo Methods*. Springer Science & Business Media, Berlin.

Moger, T. and Dhadbanjan, T. (2015). A novel index for identification of weak nodes for reactive compensation to improve voltage stability. *IET Generation, Transmission & Distribution*, 9(14):1826–1834.

Moger, T. and Dhadbanjan, T. (2016). Fuzzy logic approach for reactive power coor- dination in grid connected wind farms to improve steady state voltage stability. *IET Renewable Power Generation*, 11(2):351–361.

Nelsen, R. B. (2007). *An Introduction to Copulas*. Springer Science & Business Media, Berlin.

Oke, O. A. and Thomas, D. W. P. (2012). Enhanced cumulant method for probabilistic power flow in systems with wind generation. In *2012 11th International Conference on Environment and Electrical Engineering*, pages 849–853.

Papaefthymiou, G. and Kurowicka, D. (2008). Using copulas for modeling stochastic dependence in power system uncertainty analysis. *IEEE Transactions on Power Systems*, 24(1):40–49.

Patra, S. and Misra, R. B. (1993). Probabilistic load flow solution using method of moments. In *1993 2nd International Conference on Advances in Power System Control, Operation and Management, APSCOM-93*, vol. 2, pages 922–934.

Prusty, B. R. and Jena, D. (2016). Combined cumulant and Gaussian mixture approximation for correlated probabilistic load flow studies: A new approach. *CSEE Journal of Power and Energy Systems*, 2(2):71–78.

Prusty, B. R. and Jena, D. (2017a). A critical review on probabilistic load flow studies in uncertainty constrained power systems with photovoltaic generation and a new approach. *Renewable and Sustainable Energy Reviews*, 69:1286–1302.

Prusty, B. R. and Jena, D. (2017b). A sensitivity matrix-based temperature-augmented probabilistic load flow study. *IEEE Transactions on Industry Applications*, 53(3):2506–2516.

Rezaee Jordehi, A. (2018). How to deal with uncertainties in electric power systems? A review. *Renewable and Sustainable Energy Reviews*, 96:145–155.

Rosenblueth, E. (1975). Point estimates for probability moments. *Proceedings of the National Academy of Sciences*, 72(10):3812–3814.

Rosenblueth, E. (1981). Two-point estimates in probability. *Applied mathematical modelling*, 5(5):329–335.

Sanabria, L. A. and Dillon, T. S. (1986). Stochastic power flow using cumulants and von mises functions. *International Journal of Electrical Power and Energy Systems*, 8:47–60.

Singh, V., Moger, T., and Jena, D. (2021). Comparative evaluation of basic probabilistic load flow methods with wind power integration. In *2020 3rd International Conference on Energy, Power and Environment: Towards Clean Energy Technologies*, pages 1–6.

Singh, V., Navada, H. G., and Shubhanga, K. (2018). Large power system stability analysis using a foss-based tool: Scilab/xcos. In *2018 20th National Power Systems Conference (NPSC)*, pages 1–6. IEEE.

Su, C.-L. (2005). Probabilistic load-flow computation using point estimate method. *IEEE Transactions on Power Systems*, 20(4):1843–1851.

Wang, X. and McDonald, J. (1994). *Modern Power System Planning*. McGraw-Hill International (UK) Limited, Maidenhead.

Wang, Y. and Pierrat, L. (1997). Simulation of three-phase voltage unbalance using correlated Gaussian random variables.

Wood, A. J., Wollenberg, B. F., and Sheblé, G. B. (2013). *Power Generation, Operation, and Control*. John Wiley & Sons, Hoboken, NJ.

Yuan, Y., Zhou, J., Ju, P., and Feuchtwang, J. (2011). Probabilistic load flow computation of a power system containing wind farms using the method of combined cumulants and gram–charlier expansion. *IET Renewable Power Generation*, 5(6):448–454.

Zimmerman, R. D. and Murillo-Sánchez, C. E. (2016) Matpower 6.0 user's manual. *Power Systems Engineering Research Center*, 9.

10 Risk Evaluation of Electricity Systems with Large Penetration of Renewable Generations

Sheng Wang
State Grid (Suzhou) City & Energy Research Institute Co., Ltd

Lalit Goel
Nanyang Technological University

Yi Ding
Zhejiang University

CONTENTS

10.1 Introduction 240
    10.1.1 Importance of Risk-Based System Analysis....................................240
    10.1.2 Modeling of the Uncertainties of Wind..240
    10.1.3 Handling the Uncertainties of Renewable Energies in the
        Operation of Power Systems...241
    10.1.4 Evaluation of System Over-Limit Risk Indices 241
    10.1.5 Contribution of This Chapter..242
10.2 Temporal and Spatial Modeling of Uncertainties.......................................243
    10.2.1 Temporal and Spatial Reliability Model of Windfarms 243
    10.2.2 Operational Reliability Models of Coupling Components 245
10.3 Modeling of Power System and Uncertainty Handling Approach 247
    10.3.1 Formulation of the Optimal Control Framework 247
    10.3.2 Dynamic Gas Network Constraints..248
10.4 System Over-Limit Risk Indices and Evaluation Procedures 250
10.5 Interpretation of Probabilistic Numerical Analysis Results.......................252
    10.5.1 Case 1: Analysis of Wind Power, Load Curtailment, and
        Gas Network 253
    10.5.2 Case 2: Temporal-Spatial Reliability Analysis................................254
    10.5.3 Case 3: Comparison of Different Wind Speed Levels......................254
10.6 Conclusions...256
References...256

DOI: 10.1201/9781003271857-10

10.1 INTRODUCTION

10.1.1 Importance of Risk-Based System Analysis

With the worldwide transition toward low-carbon and sustainable energy utilization in recent years, the share of renewable generations is increasing rapidly in electricity systems. Till the end of 2018, the renewable generation (including hydro, wind, solar, biofuels, waste, geothermal, and tide) had reached 6879 TWh, an increase of 21.23% compared with 2015 [1]. In over 20 countries, the share of renewable generation exceeds 50%. A roadmap to complete renewable electricity in Europe and North Africa was proposed in [2]. In Denmark, the total wind generation equaled 47% of its electricity consumption in 2019, which continues to grow in recent years [3]. The US National Renewable Energy Laboratory proposed a blueprint that aims to realize 80% renewable electricity generation in 2050 [4]. China proposed carbon neutrality and peak carbon dioxide emission schemes, and focused on renewable generations as one of the solutions.

However, due to the fluctuation, intermittency, and unpredictability of the renewable generations, it also brings potential risks to the reliable and secure operation of electricity systems. In Denmark, it was reported that more than half of the power system's imbalanced situation was caused by the fluctuation of wind, and this kind of circumstances will occur more frequently in the future [5]. The random failure of the wind turbine itself may also become an important factor. In 2019, the simultaneous failures of Hornsea offshore windfarm and Little Barford gas-fired units (GFU) led to the blackout in the UK [6]. Therefore, it is important to study the system risks under the high penetration of wind.

10.1.2 Modeling of the Uncertainties of Wind

In order to analyze the system's risks under the large penetration of wind, the prerequisite is the modeling of wind uncertainties. The prediction method of wind speed is categorized according to different standards, as illustrated in Figure 10.1. By timescale, it could be categorized into long term, medium term, short term, and ultra-short term. According to the different mathematical methods, it can be divided into mechanism- and data-driven approaches. According to different spatial scales, the prediction can be targeted on one single wind turbine, or a windfarm, or a cluster of adjacent windfarms. According to the output results, it can also be divided into point-by-point and probabilistic predictions.

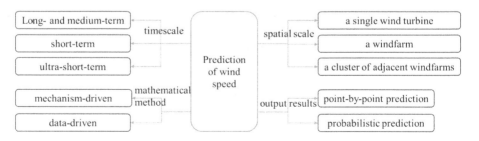

FIGURE 10.1 Systematic interpretation of wind prediction methods.

Risk Evaluation of Electricity Systems

The prediction of wind speed has been extensively studied in previous studies. A joint particle swarm optimization and gray method was used in the wind forecast in [7] to improve the accuracy. A pattern-based forecast method was proposed in [8] based on the unified key component evaluation. A multitime step-based operational wind generation forecast method was proposed in [9].

Besides the fluctuations in wind speed, the uncertainty of wind generation also comes from the random failure and repair of wind turbines. These two factors should be considered together. A reliability model of a windfarm was proposed in [10], considering both wind fluctuations and forced outages of wind turbines. The reliability of windfarm was studied in [11], which uses multistate Markov and cross-correlation function to represent the fluctuation of wind in several locations.

10.1.3 HANDLING THE UNCERTAINTIES OF RENEWABLE ENERGIES IN THE OPERATION OF POWER SYSTEMS

Many research papers have incorporated the uncertainties of renewable energies in the economic dispatch, scheduling, or planning of electricity systems. A comprehensive method for the probabilistic electric power flow was proposed in [12] to assess the fluctuation from photovoltaic generations on the electricity system operation. The simplified day-ahead scheduling method was proposed in [13] for wind generating units. A new scheduling method for generating units was established in [14] for promoting the reliability of electricity systems under high wind penetration. A risk-oriented stochastic optimization model for the capacity market was developed in [15] to minimize the maximum regret with renewable energies, and a practical case in China was investigated. Reliability-based expansion planning of electricity systems with renewable generations was conducted in [16] for lowering carbon emission.

On the other hand, a new rising technology, power-to-gas (P2G), offers a promising solution to the promotion the utilization of renewable energies. It can convert surplus electric generation from renewable generators into hydrogen or synthetic natural gas. By this means, the electricity can be indirectly stored for later use, thus preventing the waste [17]. Though the gas network does not require a precisely balanced supply and demand, excessive injection of gas into the gas network could still cause pressure fluctuations, threatening its secure operation. Under this new circumstance, the support from the gas system on renewable energy utilization and its influence on operational risks of electricity systems is worth studying. Some quantitative studies have been conducted to assess economic influence of P2Gs on the electricity systems and gas networks in the operation [18,19]. However, the impacts on system risk have not been studied yet, especially in terms of spatial and temporal risks.

10.1.4 EVALUATION OF SYSTEM OVER-LIMIT RISK INDICES

The risk evaluation of electricity systems with large penetration of renewable energies has been extensively studied in previous researches. Both long- and medium-term reliability indices were assessed in [20,21] using universal generating function techniques, respectively, with a large share of wind generation. The impacts of different wind power forecasting techniques on the risk evaluation of electricity distribution networks were evaluated in [22]. The risk of unit commitment in wind integrated electricity systems was assessed in [23].

However, the studies described above focused on analyzing the risk in the long term over the whole system, while not considering the short-term and location-varying risks in the operational phase. The powers of renewable generations, such as wind and solar, vary in time and location [24]. For example, the photovoltaic generation is highly correlated with the light intensity, which is different throughout the day. The resource endowment of wind generation also varies from coastal mountain areas. This will result in different power flow patterns in the transmission system at different time points and further affect the risk in electricity systems [25]. Therefore, spatial-temporal risk analysis is urgently required.

The temporal risk, also known as short-term or operational reliability, has been previously studied in electricity systems. Reference [26] introduced the basic concept of operational reliability. It has been demonstrated as valuable information in risk-based scheduling, load shedding devising, and other decision-making processes. The short-term risk indices, which reflect the dynamic security of renewable generations, were proposed in [27]. The spatial risk is usually specified in nodal scales in the electricity system risk assessment. It helps nodal customers to estimate the reliability and quality of their electricity supply. The correlations between nodal energy price and nodal reliability indices were studied in [28]. The nodal reliability impact from electric vehicles on electricity systems was investigated in [29]. However, few studies have evaluated the spatial and temporal risks jointly or studied the impact of renewable generations on the electricity system risks in these two dimensions.

10.1.5 CONTRIBUTION OF THIS CHAPTER

To bridge the research gaps, this chapter proposes a spatial-temporal risk assessment approach in electricity systems under large penetrations of renewable generations. The contributions of the chapter are as follows:

1. The temporal and spatial reliability of the windfarm is modeled. Both spatial and temporal correlations in the wind speed prediction and the impact of wind turbine failures are jointly considered. Moreover, the multistate reliability models of coupling components, i.e., GFU and P2G, are modeled considering the interdependency between the electricity and gas systems.
2. An optimal control framework of the electricity system under the uncertainty of wind and integration of gas systems is proposed. To avoid the potential risks on the gas system operation by the gas injection of P2Gs, the constraints imposed by gas flow dynamics are incorporated in the optimal control framework.
3. The risk indices are extended in both time and space dimensions. To this end, the impact of renewables on the electricity system operation at different locations can be quantified. Furthermore, to reduce the computation burdens, the discretized partial derivative equations (PDEs) of gas flow dynamics are further relaxed into second-order cone (SOC) constraints, so that the off-the-shelf solvers can be applied more efficiently.

Risk Evaluation of Electricity Systems 243

10.2 TEMPORAL AND SPATIAL MODELING OF UNCERTAINTIES

To study the risks to the electricity system during operation, the stochastic process of the available generating capacity of components, including the windfarm, GFU, P2G, and other traditional fossil generating units (TFU), should be modeled. Therefore, the temporal and spatial reliability of multiple windfarms is modeled, and the reliability models of GFUs and P2Gs are established.

10.2.1 TEMPORAL AND SPATIAL RELIABILITY MODEL OF WINDFARMS

The reliability of the windfarm is indicated by its available capacity to generate electricity [30]. It is related to two major factors, wind velocity and state of wind turbines. Wind velocity essentially depends on the weather system. It not only evolves in time, but is also correlated among different locations. Therefore, the precision will be significantly improved if we forecast the wind speed in both time and space dimensions.

In the operational phase, the Markov process is commonly adopted to approximate the chronological wind speed at a single location [5,21]. To establish the Markov process, the wind speed is first clustered into a finite set of states [31]. The number of states is flexible, depending on the accuracy requirement. Assume that wind speeds are the same for different wind turbines at the same windfarm. Generally, denote the number of states as NH_j^w at the windfarm j. During the operation, the wind speed takes random values from $\left\{ v_j^1, v_j^2, \ldots, v_j^{NH_j^w} \right\}$. Suppose the wind speed was at state h_0 at $t = 0$. The time-varying probability of wind speed, $\Pr^h(t)$, at each state can be calculated by solving [21]:

$$
\begin{cases}
\dfrac{d\,\Pr^h(t)}{dt} = -\Pr^h(t) \sum_{h'=1}^{NH_j^w, h' \neq h} \lambda_{h,h'}^w + \sum_{h'=1}^{NH_j^w, h' \neq h} \Pr^{h'}(t) \lambda_{h',h}^w, h = 1, 2, \ldots, NH_j^w \\[4mm]
\Pr^1 \big|_{t=0} = 0, \ldots, \Pr^{h_0} \big|_{t=0} = 1, \ldots = \Pr^{NH_j^w} \big|_{t=0} = 0
\end{cases}
\tag{10.1}
$$

where $\lambda_{h,h'}^w$ is the state transition rate from state h to h'.

To characterize the wind speed correlations in space, its concept is further extended into the temporal-spatial Markov process, as illustrated in Figure 10.2. The spatial Markov process was used in geostatistical, data processing, and image processing studies based on random field theories [32–35]. However, its applications in wind forecasting are still in their infancy. A spatial-temporal Markov process model was developed in [36] for ultra-short-term wind speed forecasting. This technique was further applied in the operational forecasting of renewable generation in [37]. To quantify the temporal and spatial effects of wind generation on the electricity system risks, it is extended to consider the random failures of wind turbines and then adopted in this chapter.

Given a dataset of wind speed, consider two windfarms j_1 and j_2. The state transition probability from windfarm j_1 at state h_1 to the windfarm j_2 at state h_2 can be calculated as

$$
\Pr_{j_1, j_2}^{h_1, h_2} = N_{j_1, j_2}^{h_1, h_2} / N_{j_1}^{h_1}
\tag{10.2}
$$

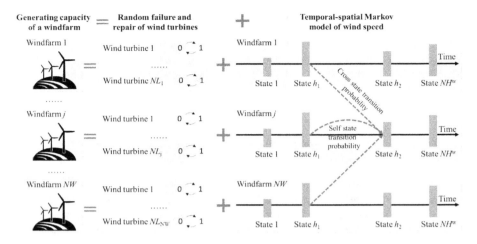

FIGURE 10.2 Temporal and spatial reliability model of windfarms.

where $N_{j_1}^{h_1}$ is the times when the state of wind speed at windfarm j_1 is h_1. $N_{j_1,j_2}^{h_1,h_2}$ is the times when the state of wind speeds at windfarms j_1 and j_2 are h_1 and h_2, respectively. When $j_1 = j_2$, it is defined as the self-transition probability as in the normal Markov process. Otherwise, when $j_1 \neq j_2$, it is defined as cross-transition probability, which describes the influence from the wind speed at other locations.

Repeat the above process until we obtain all the state transition probabilities. Then the state transition matrix between any two windfarms can be described as

$$\Pr_{j_1,j_2} = \begin{bmatrix} \Pr_{j_1,j_2}^{1,2} & \Pr_{j_1,j_2}^{1,2} & \cdots & \Pr_{j_1,j_2}^{1,NH_{j_2}^w} \\ \Pr_{j_1,j_2}^{2,1} & \Pr_{j_1,j_2}^{2,2} & \cdots & \Pr_{j_1,j_2}^{2,NH_{j_2}^w} \\ \cdots & \cdots & \cdots & \cdots \\ \Pr_{j_1,j_2}^{NH_{j_1}^w,1} & \Pr_{j_1,j_2}^{NH_{j_1}^w,2} & \cdots & \Pr_{j_1,j_2}^{NH_{j_1}^w,NH_{j_2}^w} \end{bmatrix} \quad (10.3)$$

After formulating the state transition matrix for any pair of windfarms and giving the initial values of wind speed at all the locations, the future state of wind speed can be forecast utilizing time-sequential Monte Carlo simulation (TSMCS) [5]. However, the predicted wind speed at one location may have multiple possible values from other windfarms at various locations. We set them as reference values. The final predicted value of wind speed at windfarm j at period k, $\hat{v}_{j,k}$, is calculated by the weighted average [36]:

$$\hat{v}_{j,k} = \kappa_{1,j} v_{1,j,k} + \cdots + \kappa_{j',j} v_{j',j,k} + \cdots + \kappa_{NW,j} v_{NW,j,k}, \sum_{j'=1}^{NW} \kappa_{j',j} = 1 \quad (10.4)$$

where $v_{j',j,k}$ is the wind speed at windfarm j predicted from windfarm j' at period k. NW is the number of windfarms. $\kappa_{j',j}$ is the weight coefficient from windfarm j' to windfarm j.

Risk Evaluation of Electricity Systems

To calculate appropriate weight coefficients, the following optimization problem is formulated to minimize the prediction error in the given dataset:

$$\underset{\kappa_{j',j}}{\text{Min}} \sum_{k \in K} \left(v_{j,k} - \hat{v}_{j,k} \right)^2 \tag{10.5}$$

where $v_{j,k}$ is the wind speed at windfarm j at period k in the given dataset. K is the set of periods.

According to the predicted wind speed, the available generating capacity of wind turbine l in windfarm j at period k, $G_{j,l,k}^{wt}$, can be calculated as [38]

$$G_{j,l,k}^{wt} = \begin{cases} 0 & ,0 \leq \hat{v}_{j,k} \leq v_{j,l}^{ci} \\ \left(A_{j,l} + B_{j,l} v_{j,k} + C_{j,l} \hat{v}_{j,k}^2 \right) G_{j,l}^{r} & ,v_{j,l}^{ci} \leq \hat{v}_{j,k} \leq v_{j,l}^{r} \\ P_{j,l}^{r} & ,v_{j,l}^{r} \leq \hat{v}_{j,k} \leq v_{j,l}^{co} \\ 0 & \hat{v}_{j,k} \geq v_{j,l}^{co} \end{cases} \tag{10.6}$$

where $v_{j,l}^{ci}$, $v_{j,l}^{r}$, and $v_{j,l}^{co}$ are the cut-in, rated, and cut-out wind speeds, respectively [39]. $A_{j,l}$, $B_{j,l}$, and $C_{j,l}$ are the parameters of the specific wind turbine. $G_{j,l}^{r}$ is the rated generating power of the wind turbine.

On the other hand, the operating state of the wind turbine is the other key factor determining its generating power. The reliability of a wind turbine can be described as the binary state. Thus, the available power generation capacity of windfarm j at period k, $G_{j,k}^{w}$, can be calculated as

$$G_{j,k}^{w} = \sum_{l=1}^{NL_j} I_{j,l}^{w} \times G_{j,l,k}^{wt} \tag{10.7}$$

where $I_{j,l}^{w}$ is a binary variable. $I_{j,l}^{w} = 1$ and 0 denote the perfect-functioning and complete-failure states, respectively. NL_j is the number of wind turbines at windfarm j.

10.2.2 Operational Reliability Models of Coupling Components

GFU and P2G are the two main categories of coupling components between the electricity and gas systems. Different from the TFUs such as coal-fired units, the operational reliability of coupling components not only depends on their inherent failure and repair, but also relies on the just-in-time supply of the energy they consume, e.g., gas for GFUs and electricity for P2Gs.

1. Operational reliability model of GFUs

   The operational reliability of GFU is represented using a multistate model. Its available electric generating capacity can be reduced partially or reduced to zero [40]. The number of states of the GFU j at bus i is represented as $NH_{i,j}^{gfu}$. Thus, during the operational phase, the electric generating capacity of the GFU by random failure and repair takes values from $\left\{ G_{i,j}^{gfu,1},...,G_{i,j}^{gfu,h},...,G_{i,j}^{gfu,NH_i^g} \right\}$.

The available generating capacity of a GFU also depends on the adequacy of the gas supply, which can be influenced by the status of natural gas networks. Thus, the actual electric generating capacity of GFU j at bus i, $DG_{i,j}^{gfu}$, can be calculated as

$$DG_{i,j}^{gfu} = \min \left\{ \begin{array}{l} G_{i,j}^{gfu,h}, \\ \left(-b_{i,j} + \sqrt{b_{i,j}^2 - 4a_{i,j}(c_{i,j} - HV \cdot w_{i,j}^{gfu})} \right) / \left(2a_{i,j} \right) \end{array} \right\} \quad (10.8)$$

where $a_{i,j}$, $b_{i,j}$, and $c_{i,j}$ are the operating parameters of the GFU. HV is the calorific value of gas [41], and $w_{i,j}^{gfu}$ is the gas injection for the GFU during the operation.

2. Operational reliability model of P2Gs

The operational reliability of a P2G facility also depends on both its inherent state transition and the electricity supply. The P2G devices work in parallel in a practical P2G facility, and their operating conditions are fully independent [42]. The reliability of each P2G device is represented as the binary state. We assume that the parameters of the P2G devices, including failure and repair rates, and capacities are identical. During the operation, the gas production capacity $W_{i,j}^{ptg}$ of P2G module j at bus i takes random values from $\left\{ 0, W_{i,j}^{ptg} \right\}$. Therefore, the gas production capacity of the whole P2G facility is

$$W_i^{ptg} = \sum_{i=1}^{NP_i} I_{i,j}^{ptg} \times W_{i,j}^{ptg} \quad (10.9)$$

where $I_{i,j}^{ptg}$ is a binary variable. $I_{i,j}^{ptg} = 1 = 1$ and 0 denote the perfect-functioning and complete-failure states, respectively, and NP_i is the number of P2G modules at bus i.

Gas-producing capacities of P2Gs are also determined by the adequacy of the electricity at the bus where the P2G is located. The electricity supply may be affected by the status of the electricity system. Thus, the gas-producing capacity of the P2G facility can be evaluated as [18]

$$DW_i^{ptg} = \min \left\{ W_i^{ptg}, g_i^{ptg} / \left(\eta_i^{ptg} H_g \right) \right\} \quad (10.10)$$

where DW_i^{ptg} is the gas-producing capacity of the P2G at bus i. g_i^{ptg} is the electricity supply for the P2G at bus i during the operation. η_i^{ptg} is the efficiency of the P2G facility.

10.3 MODELING OF POWER SYSTEM AND UNCERTAINTY HANDLING APPROACH

During operation, failures of system components and fluctuation of wind power could occur, which might lead to the total power generation becoming inadequate for the demand. The generators or their reserve margins will be re-dispatched. In an even worse case, load curtailment might also be implemented to maintain a balanced operation. In particular, the natural gas system also participates in the re-dispatch process to back up the electricity system through GFUs, as well as to consume the surplus wind energy through P2Gs.

10.3.1 FORMULATION OF THE OPTIMAL CONTROL FRAMEWORK

The goal of the optimal dispatch is to optimize the generation and load curtailment costs. This economic loss of load curtailment can be implicitly quantified using customer damage functions [43]. Owing to the time-dependent characteristics of gas flow dynamics, the re-dispatch is optimized over a certain period, rather than a single time point in the traditional optimal dispatch. Therefore, the optimal re-dispatch is formulated as an optimal control problem:

$$
\underset{g_{i,j,k}, g_{i,j,k}^{gfu}, w_{i,k}^{ptg}, lc_{i,k}}{\text{Min}} \quad J = \sum_{k \in K} \left(\sum_{i \in GB} \left(\rho_{i,k}^{g} \left(\sum_{j \in NG_i^{gfu}} g_{i,j,k}^{gfu} / H_g - w_{i,k}^{ptg} \right) \right) \right.
$$

$$
\left. + \sum_{i \in EB} \left(\text{CDF}_i \, lc_{i,k} + \sum_{j \in NG_i} cst_{i,j} \left(g_{i,j,k} \right) \right) \right) \tag{10.11}
$$

Subject to (12)–(18), and constraints from the gas network:

$$
G_{i,j}^{tfu,\min} \le g_{i,j,k} \le G_{i,j}^{tfu} \tag{10.12}
$$

$$
DG_{i,j}^{gfu,\min} \le g_{i,j,k}^{gfu} \le DG_{i,j}^{gfu} \tag{10.13}
$$

$$
0 \le g_{i,j,k}^{w} \le G_{i,j,k}^{w} \tag{10.14}
$$

$$
0 \le w_{i,k}^{ptg} \le DW_i^{ptg} \tag{10.15}
$$

$$
\sum_{j \in NG_i^{gfu}} g_{i,j,k}^{gfu} + \sum_{j \in NG_i} g_{i,j,k} + \sum_{j \in NW_i} g_{i,j,k}^{w} - g_{i,k}^{ptg} - D_i^{e} - \sum_{j \in \Omega_i^{e}} f_{ij,k} = 0 \tag{10.16}
$$

$$
f_{ij,k} = \left(\theta_{i,k} - \theta_{i,k} \right) / X_{ij} \tag{10.17}
$$

$$
\left| f_{ij,k} \right| \le f_{ij}^{\max} \tag{10.18}
$$

where $w_{i,k}^{ptg}$ and $g_{i,k}^{ptg}$ are the gas production and electricity consumption, respectively, of the P2G j at bus i at period k, and $\rho_{i,k}^g$ is the gas production price. $g_{i,j,k}$ is the electric output of TFU j at period k. $cst_{i,j}$ is the generation cost function for TFU. GB and EB are the sets of gas and electricity buses. NG_i and NG_i^{gfu} are the sets of TFU and GFU at bus i, respectively. $G_{i,j}^{tfu,\min}$ and $DG_{i,j}^{gfu,\min}$ are the lower limits for the electric generations of TFU and GFU, respectively. $g_{i,j,k}^w$ is the electric generation of windfarm. NW_i is the number of windfarms at bus i. D_i^e is the electricity demand at bus i. Ω_i^e is the set of electricity branches linked to bus i. $f_{ij,k}$ is the electric flow from bus i to j. $\theta_{i,k}$ is the phase angle. X_{ij} is the reactance of the electricity branch ij. $f_{ij}^{\max}$ is the capacity of electricity line ij. The constraints from the gas network are more complicated, which will be discussed in the next section.

10.3.2 DYNAMIC GAS NETWORK CONSTRAINTS

The idea of imposing constraints from the gas network is to limit the gas withdrawals and injections of GFUs and P2Gs, to ensure the secure operation of the gas network.

1. Reformulation of gas flow dynamic equations

    Two PDEs, namely continuity and motion equations, govern the dynamics of gas flow in a pipeline. In a horizontal gas pipeline, the dissipative and isothermal gas flow is described by [44]:

$$\frac{B^2}{\rho_0 A}\frac{\partial q}{\partial x} + \frac{\partial p}{\partial t} = 0 \tag{10.19}$$

$$\frac{\partial p}{\partial x} + \frac{\rho_0}{A}\frac{\partial q}{\partial t} + \frac{2\rho_0^2 B^2}{F^2 DA^2}\frac{q|q|}{p} = 0 \tag{10.20}$$

    where q and p are the quantity of gas flow and gas pressure, respectively. A is the cross-section area of the pipeline. B is the wave speed of natural gas. ρ_0 is the gas density at the standard temperature and pressure. D is the diameter. F is Fanning transmission coefficient.

    The above PDEs for the pipeline ij can be discretized using the Wendroff formula [18]:

$$\frac{1}{B^2}\left(p_{m+1,k+1} + p_{m,k+1} - p_{m+1,k} - p_{m,k}\right) + \frac{\Delta t\rho_0}{\Delta xA}\left(q_{m+1,k+1} - q_{m,k+1} + q_{m+1,k} - q_{m,k}\right) = 0 \tag{10.21}$$

$$4\left(\left(p_{m+1,k+1} + p_{m+1,k}\right)^2 - \left(p_{m,k+1} + p_{m,k}\right)^2\right)$$

$$+ \frac{\psi\rho_0 B^2 \Delta x}{F^2 DA^2}\left(q_{m+1,k+1} + q_{m+1,k} + q_{m,k+1} + q_{m,k}\right)^2 = 0 \tag{10.22}$$

$$\operatorname{sgn}(x) = \begin{cases} 1, x \geq 0 \\ -1, x < 0 \end{cases} \tag{10.23}$$

where Δx and Δt are the step sizes in length and time domains. m is the index of pipeline sections. ψ represents the direction of the gas flow, where $\psi = \mathrm{sgn}(p_i - p_j)$. $\mathrm{sgn}(x)$ is the sign function defined by (10.23).

Assume that the gas flow does not change direction during the operation [45]. Then, (10.22) can be further relaxed into SOC constraints:

$$\left\| p_{m,k+1} + p_{m,k}, \sqrt{\frac{\rho_0^2 B^2}{F^2 DA^2}} \Delta x \left(q_{m+1,k+1} + q_{m+1,k} + q_{m,k+1} + q_{m,k} \right) \right\|_2 (1+\psi) +$$

$$\left\| p_{m+1,k+1} + p_{m+1,k}, \sqrt{\frac{\rho_0^2 B^2}{F^2 DA^2}} \Delta x \left(q_{m+1,k+1} + q_{m+1,k} + q_{m,k+1} + q_{m,k} \right) \right\|_2 (1-\psi)$$

$$\leq (1+\psi)\left(p_{m+1,k+1} + p_{m+1,k} \right) + (1-\psi)\left(p_{m,k+1} + p_{m,k} \right) \tag{10.24}$$

3. Initial and boundary conditions

The initial and boundary conditions of the PDEs of the gas flow dynamics are specified in this subsection. The initial condition is given by the state variables of the electricity and gas systems in the normal state, where only the calculation of the gas flow in the steady-state is involved. The coordination of electricity and gas systems aims to minimize operating cost C_{IEGS} by controlling the generation schedule of TFUs and GFUs and the gas production schedule of gas sources:

$$\min_{w_i, g_{i,j}, g_{i,j}^{gfu}} C_{IEGS} = \sum_{i \in GB} \rho_i^g w_i + \sum_{i \in EB} \sum_{j \in NG_i} cst_{i,j}\left(g_{i,j} \right) \tag{10.25}$$

Subject to (10.26)–(10.29) and the constraints for electricity system (10.12)–(10.18):

$$w_i^{\min} \leq w_i \leq w_i^{\max} \tag{10.26}$$

$$w_i - D_i^g - \sum_{j \in NG_i^{gfu}} w_{i,j}^{gfu} + w_i^{ptg} - \sum_{j \in \Omega_i^g} q_{ij} = 0 \tag{10.27}$$

$$q_{ij} = C_{ij} \Gamma_{ij} \sqrt{\left| p_i^2 - p_j^2 \right|} \tag{10.28}$$

$$\left| q_{ij} \right| \leq q_{ij}^{\max} \tag{10.29}$$

where w_i is the gas produced by the gas well at bus i. $w_i^{\min}$ and $w_i^{\max}$ are the minimum and maximum gas productions of the gas source. Ω_i^g represents the set of gas pipelines that are connected to bus i. q_{ij} denotes the gas flow from bus i to j. C_{ij} is a characteristic parameter of the pipeline, depending on the length, absolute rugosity, and some other properties. $q_{ij}^{\max}$ denotes the maximum transmission capability in the gas pipeline ij.

250 Renewable Energy Integration to the Grid

After solving the optimization problem above, the values of w_i, $g_{i,j}$, and $g_{i,j}^{gfu}$ can be obtained. During the intraday operation, the gas pressures in the pipeline segment m in period k, $p_{ij,m,k}$, should be controlled within the secure limits:

$$(1-\gamma)p_{ij,m} \leq p_{ij,m,k} \leq (1+\gamma)p_{ij,m} \tag{10.30}$$

where γ is the tolerance of gas pressure. The initial conditions for the gas pressure and gas flow are specified as

$$p_{ij,m,0} = \sqrt{p_i^2 - \Gamma_{ij}q_{ij}^2\left(C_{ij}^2 L_{ij}\right)^{-1} m\Delta x} \tag{10.31}$$

$$q_{ij,m,0} = q_{ij} \tag{10.32}$$

where L_{ij} is the length of the pipeline ij.

In the gas network, the boundary conditions among the pipelines are

$$p_{ij,0,k} = p_{ij_1,0,k}\left(\forall j_1 \in \Omega_i^g, \forall k\right) \tag{10.33}$$

$$p_{ij,0,k} = p_{j_2i,M_{ij},k}\left(\forall j_2 \in \Omega_i^g, \forall k\right) \tag{10.34}$$

$$w_i - D_i^g + w_i^{ptg} - \sum_{j=1}^{NG_i^{gfu}} w_{i,j}^{gfu} + \sum_{j\in\Omega_i^g} q_{ji,M_{ij},k} - \sum_{j\in\Omega_i^g} q_{ij,0,k} = 0, \forall k \tag{10.35}$$

where M_{ij} is the number of pipeline segments in pipeline ij.

10.4 SYSTEM OVER-LIMIT RISK INDICES AND EVALUATION PROCEDURES

The spatial-temporal risk evaluation of the electricity system is the process of predicting risks for the system operator and nodal customers under a given system operating condition. TSMCS is used to sample the temporal and spatial wind speed, chronological random failure, and repair of generators and P2Gs during the operation and calculate the risk indices. The expected interruption of demand (EID) and risk of system overload (RSOL) of the electricity system are evaluated, as in (10.36) and (10.37) [23, 27]. It is worth mentioning that these two risk indices are extended to time-varying indices and are further specified into nodal scale, which provides better flexibility to reflect the impact of the temporal and spatial correlations of the renewables on the operational risks of the electricity system.

$$\text{EID}_i(t) = \left(\sum_{n=1}^{NS} \int_0^t lc_i(\tau)d\tau\right) / NS \tag{10.36}$$

Risk Evaluation of Electricity Systems 251

$$\text{RSOL}_i(t) = \left(\sum_{n=1}^{NS} \int_0^t \text{flag}\big(lc_i(\tau)\big) d\tau \right) / NS \tag{10.37}$$

$$\text{flag}(x) = \begin{cases} 1, x > 0 \\ 0, x \le 0 \end{cases} \tag{10.38}$$

where n and NS are the index and numbers of simulation times, respectively. flag(x) is the flag function as defined in (10.38). The stopping criterion for the TSMCS is calculated by [46]

$$\max \left\{ \sqrt{\text{Var}\left(\sum_{i \in EB} \text{EID}_i(t) \right)} \Big/ \sum_{i \in EB} \text{EID}_i(t), \sqrt{\text{Var}\left(\sum_{i \in EB} \text{RSOL}_i(t) \right)} \Big/ \sum_{i \in EB} \text{RSOL}_i(t) \right\} \le \xi \tag{10.39}$$

where Var represents the variance function.

The temporal-spatial reliability evaluation procedure for the electricity system with renewable generation is presented with the following steps:

Step 1: Input wind speed data. Choose a random initial state for wind speed. Set the initial states of wind turbines, generators, and P2Gs to the perfect-functioning state.

Step 2: Calculate the initial operating condition of the gas system by solving the optimization problem in (10.12)–(10.18) and (10.25)–(10.29). Set the initial gas flow and pressure in the gas pipelines according to (10.31) and (10.32).

Step 3: Cluster the wind speed at each windfarm into NH_j^w states from the wind speed database. Calculate the state transition matrix between any pair of windfarms according to (10.2) and (10.3).

Step 4: Simulate the temporal and spatial wind speed according to (10.4). Compare the predicted value and the actual value in the wind speed database, and calculate the optimal weight coefficients by minimizing the prediction error, as in (10.5).

Step 5: Predict the wind speed at each windfarm with the given weight coefficients.

Step 6: Simulate the random failure and repair of wind turbines using TSMCS. Generate the available generating capacity sequence of windfarms according to (10.6) and (10.7).

Step 7: Simulate the random failures and repairs of GFUs, P2Gs, and TFUs according to section 10.2.2.

Step 8: Solve the optimal control of the electricity system considering the dynamics of gas flow, according to (10.11)–(10.18), (10.21), (10.24), (10.30), and (10.33)–(10.35), and then obtain the electric load curtailment.

Step 9: Calculate the temporal-spatial risk indices according to (10.36) and (10.37). Evaluate the stopping criterion for the TSMCS as in (10.39). If the criterion is satisfied, then output the risk indices as the final results. Otherwise, repeat the next simulation from Step 4.

10.5 INTERPRETATION OF PROBABILISTIC NUMERICAL ANALYSIS RESULTS

To verify the proposed temporal and spatial risk assessment technique, an IEEE 24-bus RTS [47] and the integration of Belgium natural gas system [48], as presented in Figure 10.3, is constructed in this section [49]. We replace the 400 MW nuclear generating unit at electric bus (EB) 18, a 100 MW unit at EB 7, and two 50 MW generators at EB 22 by windfarms of the same generating capacities. The model type and parameters of the wind turbine are set according to [30]. The oil steam generating units at EB 2, 13, 14, and 15 are replaced by the GFUs with the same capacity. Three P2Gs are installed at the gas bus (GB) 7, 10, and 16, and their capacities are set according to [49]. Other parameters including GFUs, P2Gs, gas prices, and heat value of gas are set according to [50]. The gas pressures are limited within 0.95–1.05 times of their values at the normal operating state. Numerical case studies were conducted on a Lenovo laptop with an 8565U CPU and 16GB RAM.

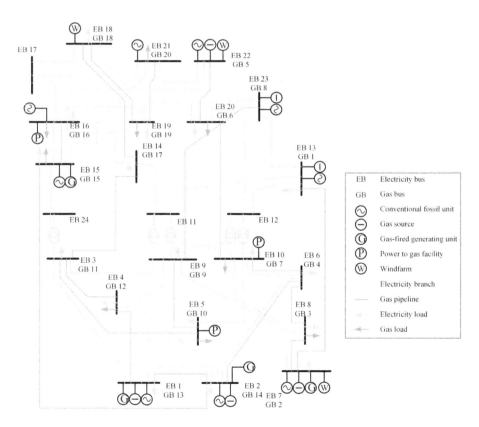

FIGURE 10.3 IEEE Reliability Test System with the integration of Belgium natural gas transmission system.

10.5.1 Case 1: Analysis of Wind Power, Load Curtailment, and Gas Network

The wind speed data were acquired from the past ten years' historical data in Texas, the US, from the National Oceanic and Atmospheric Administration (NOAA). Four locations were compared, as in Figure 10.4. The probability density of the wind speeds in four areas presents similar patterns. This indicates that there may exist spatial correlations. We further cluster the wind speed into eight states for Monte Carlo simulation.

During simulation, one representative scenario is presented to elaborate on the effect of the large share of wind generation on the electricity system operation, and how the fluctuation of wind power influences the gas system through P2Gs.

As shown in Figure 10.5, the large share of wind generation could endanger the operation of electricity system operation, even if all other generating units function perfectly. At 3:00 and 7:00 before the 400 MW unit failure, there are small electric load curtailments of 0.25 and 0.89 MW, respectively. After the unit failure at 11:00, though the wind power is approximately the same during 12:00–16:00 as in 3:00–7:00, the electric load curtailment increases dramatically. The total electric

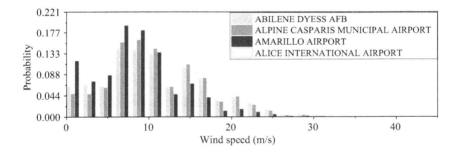

FIGURE 10.4 Histogram of wind speed data in Texas.

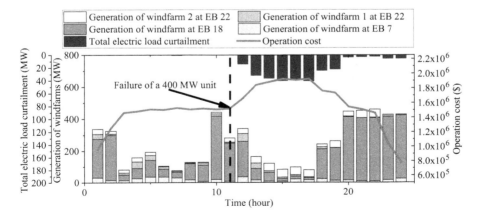

FIGURE 10.5 Wind generations, load curtailment, and operation cost in a representative scenario.

energy loss reaches 269 MWh. Due to this, the operation cost also increases. This indicates that with this level of wind penetration, the electricity system is vulnerable against possible failures of large generating units.

From the perspective of the gas system, as in Figure 10.6, we can observe that the gas system provides effective support to the electricity system. During the wind power's peak period, P2Gs consumed surplus electricity and produced 0.13 Mm^3 gas. While during the electricity generation shortage, the GFU raises its production by 24.16% to cover the electric load. Although the gas injection from P2Gs leads to slight fluctuations in the nodal gas pressures, they are still within a secure and controllable range.

10.5.2 Case 2: Temporal-Spatial Reliability Analysis

The temporal-spatial reliability indices are calculated in this case. EID and RSOL of the whole electricity system during the operational phase are shown in Figure 10.7. All components are preset in perfect-functioning states, so the system EID and RSOL are zero at the beginning. With the state transitions of wind speed and other generators, EID and RSOL increase to 2.58 MW and 0.023, respectively.

EID and RSOL during the operational phase are further specified to each EB, as presented in Figures 10.8 and 10.9, respectively. We can observe that EB 10, 9, and 5 have the highest EID and RSOL. Particularly for EB 10, its RSOL is significantly higher than that of other EBs, which indicates that those EBs are more likely to suffer load curtailment.

10.5.3 Case 3: Comparison of Different Wind Speed Levels

The endowment of wind power varies in different locations in the world. To draw a more generalized conclusion, this section conducts a sensitivity analysis, to explore the effect of different wind speed levels on temporal-spatial risks of the

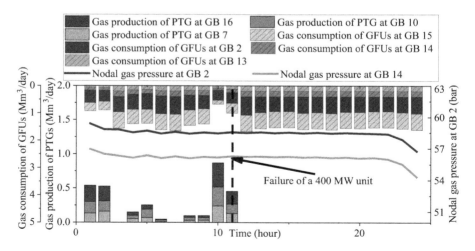

FIGURE 10.6 Gas production of P2Gs, gas consumption of GFUs, and nodal gas pressure in a representative scenario.

Risk Evaluation of Electricity Systems

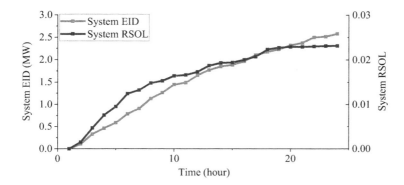

FIGURE 10.7 EID and RSOL of the electricity system.

electricity system. The probability distribution of wind speed is the same as in Case 1, while the expectation of the stochastic wind speed is different.

Figure 10.10 shows the variation of wind speed expectations with the system-level risk indices. With the increase in the expectation of wind speed, the system risk indices first decrease and then increase, which is due to the cut-out speed of the wind turbines. The lowest risks appear around 13.5–15 m/s.; above this speed zone, with

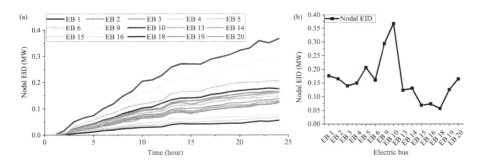

FIGURE 10.8 (a) Nodal EID during the operational phase. (b) Nodal EID at $t=24$ h.

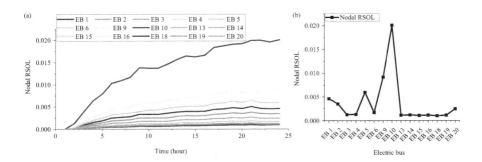

FIGURE 10.9 (a) Nodal RSOL during the operational phase. (b) Nodal RSOL at $t=24$ h.

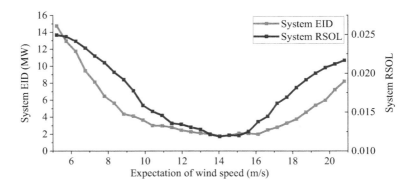

FIGURE 10.10 Sensitivity analysis of system risks with respect to the expectation of wind speed.

the increase of wind speed expectation, the wind speed will be more likely to exceed the cut-out speed, and the generation of windfarms will decrease. Thus, the electricity system becomes more likely to suffer a load curtailment.

10.6 CONCLUSIONS

The temporal-spatial risks of electricity systems under a large share of wind power have been evaluated in this chapter. Natural gas networks are integrated to the electricity system to promote the utilization of wind power. Numerical simulations indicate that the large share of renewable generations does have influences on the reliability of the electricity system. Its fluctuation, as well as intermittency, makes electricity systems more vulnerable against possible generating unit failures, e.g., the electric load curtailment even increased to 40.53 MW. On the other hand, the gas system provides effective support to the electricity system. During the wind generation's peak period, P2Gs consume electricity to produce 0.13 Mm^3 gas, while during the electricity generation shortage, the GFU raises its production by 24.16% to satisfy the electric load. Furthermore, the temporal and spatial risk indices indicate that the risk of the system is increasing during the operation. EB 10 is most likely to suffer from load curtailment.

The conclusions strongly indicate the necessity of temporal and spatial risk evaluation when the electricity systems are highly penetrated with renewable energies. The risk assessment technique proposed in this chapter can be further developed and utilized in contingency management, day-ahead unit commitment, and other pertinent decision-making.

REFERENCES

1. IEA, "Data and statistics," International Energy Agency, 2020.
2. G. Schellekens, A. Battaglini, J. Lilliestam, et al., "100% renewable electricity: A roadmap to 2050 for Europe and North Africa," 2010.
3. J. Gronholt-Pedersen, "Denmark sources record 47% of power from wind in 2019," A. Richardson and K. Donovan, eds., 2020.

Risk Evaluation of Electricity Systems **257**

4. T. Mai, D. Sandor, R. Wiser, et al., *Renewable electricity futures study. Executive summary*, National Renewable Energy Lab. (NREL), Golden, CO (United States), 2012.

5. Y. Ding, L. Cheng, Y. Zhang, et al., "Operational reliability evaluation of restructured power systems with wind power penetration utilizing reliability network equivalent and time-sequential simulation approaches," *Journal of Modern Power Systems and Clean Energy*, vol. 2, no. 4, pp. 329–340, Dec 2014.

6. J. Ambrose, "What are the questions raised by the UK's recent blackout?", 2019.

7. Y. Zhang, H. Sun, and Y. Guo, "Wind power prediction based on PSO-SVR and grey combination model," *IEEE Access*, vol. 7, pp. 136254–136267, 2019.

8. Q. Hu, P. Su, D. Yu, et al., "Pattern-based wind speed prediction based on generalized principal component analysis," *IEEE Transactions on Sustainable Energy*, vol. 5, no. 3, pp. 866–874, 2014.

9. N. Safari, C. Y. Chung, and G. C. D. Price, "Novel multi-step short-term wind power prediction framework based on chaotic time series analysis and singular spectrum analysis," *IEEE Transactions on Power Systems*, vol. 33, no. 1, pp. 590–601, 2018.

10. S. Sulaeman, M. Benidris, J. Mitra, et al., "A wind farm reliability model considering both wind variability and turbine forced outages," *IEEE Transactions on Sustainable Energy*, vol. 8, no. 2, pp. 629–637, 2017.

11. N. Nguyen, S. Almasabi, and J. Mitra, "Impact of correlation between wind speed and turbine availability on wind farm reliability," *IEEE Transactions on Industry Applications*, vol. 55, no. 3, pp. 2392–2400, 2019.

12. B. R. Prusty, and D. Jena, "A sensitivity matrix-based temperature-augmented probabilistic load flow study," *IEEE Transactions on Industry Applications*, vol. 53, no. 3, pp. 2506–2516, 2017.

13. R. Karki, S. Thapa, and R. Billinton, "A simplified risk-based method for short-term wind power commitment," *IEEE Transactions on Sustainable Energy*, vol. 3, no. 3, pp. 498–505, 2012.

14. M. Fan, K. Sun, D. Lane, et al., "A novel generation rescheduling algorithm to improve power system reliability with high renewable energy penetration," *IEEE Transactions on Power Systems*, vol. 33, no. 3, pp. 3349–3357, Feb 2018.

15. R. Lu, T. Ding, B. Qin, et al., "Reliability based min–max regret stochastic optimization model for capacity market with renewable energy and practice in China," *IEEE Transactions on Sustainable Energy*, vol. 10, no. 4, pp. 2065–2074, Oct 2019.

16. A. Moreira, D. Pozo, A. Street, et al., "Reliable renewable generation and transmission expansion planning: Co-optimizing system's resources for meeting renewable targets," *IEEE Transactions on Power Systems*, vol. 32, no. 4, pp. 3246–3257, Jul 2017.

17. J. Yang, N. Zhang, Y. Cheng, et al., "Modeling the operation mechanism of combined P2G and gas-fired plant with CO_2 recycling," *IEEE Transactions on Smart Grid*, vol. 10, no. 1, pp. 1111–1121, Jan 2019.

18. S. Clegg, and P. Mancarella, "Integrated modeling and assessment of the operational impact of power-to-gas (P2G) on electrical and gas transmission networks," *IEEE Transactions on Sustainable Energy*, vol. 6, no. 4, pp. 1234–1244, May 2015.

19. S. Chen, Z. Wei, G. Sun, et al., "Multi-linear probabilistic energy flow analysis of integrated electrical and natural-gas systems," *IEEE Transactions on Power Systems*, vol. 32, no. 3, pp. 1970–1979, May 2017.

20. Y. Ding, P. Wang, L. Goel, et al., "Long-term reserve expansion of power systems with high wind power penetration using universal generating function methods," *IEEE Transactions on Power Systems*, vol. 26, no. 2, pp. 766–774, Aug 2011.

21. Y. Ding, C. Singh, L. Goel, et al., "Short-term and medium-term reliability evaluation for power systems with high penetration of wind power," *IEEE Transactions on Sustainable Energy*, vol. 5, no. 3, pp. 896–906, Jul 2014.

22. M. Al-Muhaini, A. Bizrah, G. Heydt, et al., "Impact of wind speed modelling on the predictive reliability assessment of wind-based microgrids," *IET Renewable Power Generation,* vol. 13, no. 15, pp. 2947–2956, Nov 2019.
23. R. Billinton, B. Karki, R. Karki et al., "Unit Commitment Risk Analysis of Wind Integrated Power Systems," *IEEE Transactions on Power Systems,* vol. 24, no. 2, pp. 930–939, May 2009.
24. B. R. Prusty, and D. Jena, "An over-limit risk assessment of PV integrated power system using probabilistic load flow based on multi-time instant uncertainty modeling," *Renewable Energy,* vol. 116, pp. 367–383, 2018.
25. B. R. Prusty, and D. Jena, "A spatiotemporal probabilistic model-based temperature-augmented probabilistic load flow considering PV generations," *International Transactions on Electrical Energy Systems,* vol. 29, no. 5, pp. e2819, 2019.
26. S. Yuanzhang, C. Lin, Y. Xiaohui, et al., "Overview of power system operational reliability," In *2010 IEEE 11th International Conference on Probabilistic Methods Applied to Power Systems,* Singapore, 2010, pp. 166–171.
27. S. Datta, and V. Vittal, "Operational risk metric for dynamic security assessment of renewable generation," *IEEE Transactions on Power Systems,* vol. 32, no. 2, pp. 1389–1399, Mar 2017.
28. P. Wang, Y. Ding, and Y. Xiao, "Technique to evaluate nodal reliability indices and nodal prices of restructured power systems," *IEE Proceedings - Generation, Transmission and Distribution,* vol. 152, no. 3, pp. 390–396, May, 2005.
29. D. Tang, and P. Wang, "nodal impact assessment and alleviation of moving electric vehicle loads: from traffic flow to power flow," *IEEE Transactions on Power Systems,* vol. 31, no. 6, pp. 4231–4242, Feb, 2016.
30. A. S. Dobakhshari, and M. Fotuhi-Firuzabad, "A reliability model of large wind farms for power system adequacy studies," *IEEE Transactions on Energy Conversion,* vol. 24, no. 3, pp. 792–801, Sep 2009.
31. R. Karki, H. Po, and R. Billinton, "A simplified wind power generation model for reliability evaluation," *IEEE Transactions on Energy Conversion,* vol. 21, no. 2, pp. 533–540, Jun, 2006.
32. F. Melgani, and S. B. Serpico, "A Markov random field approach to spatio-temporal contextual image classification," *IEEE Transactions on Geoscience and Remote Sensing,* vol. 41, no. 11, pp. 2478–2487, Nov 2003.
33. R. Lowe, T. C. Bailey, D. B. Stephenson et al., "Spatio-temporal modelling of climate-sensitive disease risk: Towards an early warning system for dengue in Brazil," *Computers & Geosciences,* vol. 37, no. 3, pp. 371–381, Mar 2011.
34. W. J. M. G. Li, "Markov chain random fields for estimation of categorical variables," *Mathematical Geology,* vol. 39, no. 3, pp. 321–335, 2007.
35. Q. Yu, G. Medioni, and I. Cohen, "Multiple target tracking using spatio-temporal markov chain monte carlo data association," In *2007 IEEE Conference on Computer Vision and Pattern Recognition,* 2007, pp. 1–8.
36. Y. Zhao, L. Ye, Z. Wang, et al., "Spatio-temporal Markov chain model for very-short-term wind power forecasting," *The Journal of Engineering,* vol. 2019, no. 18, pp. 5018–5022, 2019.
37. M. He, L. Yang, J. Zhang, et al., "A spatio-temporal analysis approach for short-term forecast of wind farm generation," *IEEE Transactions on Power Systems,* vol. 29, no. 4, pp. 1611–1622, Jul 2014.
38. P. Giorsetto, and K. F. Utsurogi, "Development of a new procedure for reliability modeling of wind turbine generators," *IEEE Transactions on Power Apparatus and Systems,* vol. PAS-102, no. 1, pp. 134–143, Jan 1983.
39. Y. Sun, Z. Li, X. Yu, et al., "Research on ultra-short-term wind power prediction considering source relevance," *IEEE Access,* vol. 8, pp. 147703–147710, 2020.

Risk Evaluation of Electricity Systems

40. M. Reshid, and M. Abd Majid, "A multi-state reliability model for a gas fueled cogenerated power plant," *Journal of Applied Science,* vol. 11, no. 11, pp. 1945–1951, 2011.

41. A. Seungwon, L. Qing, and T. W. Gedra, "Natural gas and electricity optimal power flow," In *2003 IEEE PES Transmission and Distribution Conference and Exposition,* Dallas, TX, USA, 2003, pp. 138–143.

42. G. Gahleitner, "Hydrogen from renewable electricity: An international review of power-to-gas pilot plants for stationary applications," *International Journal of Hydrogen Energy,* vol. 38, no. 5, pp. 2039–2061, Feb 2013.

43. G. Wacker, and R. Billinton, "Customer cost of electric service interruptions," *Proceedings of the IEEE,* vol. 77, no. 6, pp. 919–930, Jun 1989.

44. I. Cameron, "Using an excel-based model for steady-state and transient simulation," In *PSIG Annual Meeting,* St. Louis, Missouri, 1999, pp. 39.

45. Y. Zhou, C. Gu, H. Wu, et al., "An equivalent model of gas networks for dynamic analysis of gas-electricity systems," *IEEE Transactions on Power Systems,* vol. 32, no. 6, pp. 4255–4264, Nov 2017.

46. H. Jia, Y. Ding, Y. Song, et al., "Operating reliability evaluation of power systems considering flexible reserve provider in demand side," *IEEE Transactions on Smart Grid,* vol. 10, no. 3, pp. 3452–3464, May 2019.

47. C. Grigg, P. Wong, P. Albrecht, et al., "The IEEE reliability test system-1996. A report prepared by the reliability test system task force of the application of probability methods subcommittee," *IEEE Transactions on Power Systems,* vol. 14, no. 3, pp. 1010–1020, Aug 1999.

48. D. De Wolf, and Y. Smeers, "The gas transmission problem solved by an extension of the simplex algorithm," *Management Science,* vol. 46, no. 11, pp. 1454–1465, 2000.

49. W. Sheng, D. Yi, Y. Chengjin, et al., "Reliability evaluation of integrated electricity–gas system utilizing network equivalent and integrated optimal power flow techniques," *Journal of Modern Power Systems and Clean Energy,* vol. 7, pp. 1523–1535, Oct 2019.

50. C. Unsihuay, J. W. M. Lima, and A. C. Zambroni de Souza, "Modeling the integrated natural gas and electricity optimal power flow," in *2007 IEEE Power Engineering Society General Meeting,* Tampa, FL, USA, 2007, pp. 1–7.

Index

active and reactive power sharing 158
active power loss 137
analytical methods 203
ancillary services 42, 111
ant lion optimization (ALO) 134
approximate methods 203
array yield 56, 57, 61

bender decomposition 173
beta distribution 201
Big-M 183, 184
binary cycle power plant 22, 24
biogeography-based optimization (BBO) 134

capacity outage probability table (COPT) 201
chance constrained programming 180
Clayton Copula 191, 192, 202
correlation 190, 202, 250
Cornish-Fisher expansion (CFE) 205
Copula function 190, 191
Copula theory 193, 201, 202
correlation coefficient matrix 190
Cuckoo search algorithm (CSA) 134
cumulant method (CM) 203, 205
cumulants 205, 206
cumulative distribution function (CDF) 190

differential evolution (DE) 133
discrete distribution 171, 201
distributed generation 30, 94, 102
doubly fed induction generator (DFIG) 113, 156
dry steam power plants 22

edgeworth series expansion (ESE) 205
energy storage system 50
equality constraints 139, 140
expansion planning 200, 241
expected interruption of demand (EID) 250

Federal Energy Regulatory Commission (FERC) 111
final yield 56, 57, 62
flash point power plants 22
flexible AC transmission system (FACTS) 110, 119, 120
flexible ramping 165, 181, 182, 183, 184
Frank Copula 191, 192, 202
frequency control ancillary services (FCAS) 112
full stochastic optimization 168
fuzzy logic controller 33, 38

gas fired unit 240
gas flow dynamics 242, 247, 249
gas system 241, 245, 249, 252
Gaussian Copula 191, 192
Gaussian mixture approximation (GMA) 205
genetic algorithm 119, 133
Gram-Charlier expansion (GCE) 205
grid integration 39, 50, 65
Gumbel Copula 191, 194, 195

harmonics distortion 31
horizontal-axis wind turbine (HAWT) 7, 155

IEC 61724 55, 68, 71, 72
IEEE 13-bus distribution system 101
IEEE 24-bus RTS 252
IEEE 30-bus system 134
IEEE 57-bus system 141
IEEE 118-bus system 134
IEEE 1547 65, 95
incident solar radiation 58
independent system operator (ISO) 112, 168
Indian southern zonal power grid 213
inequality constraints 115, 119, 136, 139
integrated power system 110, 120
integration of wind power 113

Kendall rank correlation coefficient (KRC) 190
kernel density estimation (KDE) 234
K-means clustering 205

L-Index 137, 141
load demand, stochastic nature of 232
load flow 81, 111, 114, 139
loading conditions 81, 109, 114
loss of load probability (LOLP) 180
low-voltage ride through (LVRT) 100

management of reactive power 108, 110, 136
management of renewable energy 96
microgrid 36, 37, 42, 53
mixed integer linear programming (MILP) 167
modeling of solar power generation 157
modeling of wind power generation 157
modern intelligent techniques 114
moments 202, 209, 210
Monte-Carlo simulation (MCS) 169, 178, 203
multistate reliability model 242

261

262   Index

National Renewable Energy Laboratory
(NREL) 240
Newton-Raphson technique 114, 139
numerical methods 203

ocean thermal energy conversion (OTEC) 31
on-load tap changer (OLTC) 80, 83, 91, 92
operational reliability model 245, 246
optimal active power dispatch (OAPD) 149, 150
optimal control 242, 247
optimal power flow (OPF) 114, 132
optimal reactive power dispatch (ORPD) 132,
141, 153
over-limit risk indices 241, 250

performance ratio 57, 68, 70, 72, 75
periodic monitoring 67
PI controller 34
pitch adjustment control method 10
point estimation method (PEM) 203, 209, 215, 216
polynomial normal transformation (PNT)
201, 202
power quality 50, 93, 96
power system planning 33, 169, 232
power-to-gas (P2G) 241
probabilistic forecasting 66
probabilistic load flow (PLF) 189, 200, 232
probabilistic steady-state analysis (PSSA) 189
product moment correlation coefficient
(PMCC) 190
PV battery systems 53

random variables (RVs) 190, 201
reactive power control 51, 66, 108, 111, 117, 118
reference yield 56, 57, 60
renewable energy generation 166, 170
renewable energy sources 39, 50, 90, 153
renewable power generation 30, 79, 153
RES Impact on voltage 90, 91
risk evaluation 239, 241, 250
risk of system overload (RSOL) 250
robust optimization 175, 176

sampling methods 203
Sklar's theorem 191
soiling level 70, 75

soiling ratio 68, 69, 75
spatial correlation 250, 253
spearman rank-order correlation coefficient
(SRCC) 190
STATCOM controllers 33
static synchronous compensators (SSC) 51
STC corrected performance ratio 72
stochastic dependence 191
sustainable development 3
sustainable energy 240
synchronous condenser 116, 117, 124

TCS Controller 33
temperature-augmented probabilistic load flow
(TPLF) 51
temperature corrected performance ratio 68, 70, 72
temporal correlation 242
thermal energy storage (TES) 12
time-sequential Monte Carlo simulation
(TSMCS) 244
total harmonic distortion 47
total voltage deviation 137, 158
traditional fossil generating units (TFU) 243

uncertainty modeling 168, 169, 200
unit commitment 166, 171, 177, 181, 242

vertical-axis wind turbine (VAWT) 7
voltage collapse 88, 90, 137
voltage collapse proximity indicator (VCPI) 90
voltage control 30, 38, 51, 80, 91, 94
voltage issues 79, 95, 97
voltage regulation 31, 38, 83, 232
voltage stability 80, 82, 88, 118

water cycle algorithm (WCA) 134
wave energy converter (WEC) 19
Weibull distribution 201, 214, 229, 234
wind turbines 96, 113, 155, 156
windfarms 240, 243
wind power
forecasting 241
uncertainties 158
wind speed, probability distribution of 255

Yaw mechanism 9, 10

CPSIA information can be obtained
at www.ICGtesting.com
Printed in the USA
BVHW091744190422
634676BV00002B/33